CROP PROTECTION MONOGRAPHS

CROP PROTECTION MONOGRAPHS

Already published volumes

Advisory Work in Crop Pest and Disease Management
Edited by J. Palti and R. Ausher

Further volumes in preparation

H. Waibel
The Economics of Integrated Pest Control in Irrigated Rice

D. P. Singh
Breeding for Resistance to Diseases and Insect Pests

I. Wahl and G. Fischbeck
Role of Wild Relatives in Control of Crop Diseases, Development of Their Epidemics, and Evolution of Pathogens

Advisory Work in Crop Pest and Disease Management

Edited by
J. Palti and R. Ausher

With Contributions by
Z. Arenstein · R. Ausher · W. Beicht · B. D. Blair
C. H. Blazquez · A. Dinoor · I. Dishon · G. Edelbaum
Y. Elkana · J. Eshel · J. T. Fletcher · H. Frankel
R. E. Frisbie · A. Genizi · J. Golan · K. Hanuss
R. Hochberg · G. M. McWhorter · J. Palti · Y. Sachs
I. J. Thomason · N. C. Toscano

With 53 Figures

Springer-Verlag
Berlin Heidelberg New York Tokyo

Dr. JOSEF PALTI
Agricultural Research Organization
The Volcani Center
P.O. Box 6
Bet Dagan 50250, Israel

Dr. REUBEN AUSHER
Extension Service
Ministry of Agriculture
P.O. Box 7054
Tel Aviv 61070, Israel

ISBN 3-540-16242-9 Springer-Verlag Berlin Heidelberg New York Tokyo
ISBN 0-387-16242-9 Springer-Verlag New York Heidelberg Berlin Tokyo

Library of Congress Cataloging-in-Publication Data. Main entry under title: Advisory work in crop pest and disease management. (Crop protection monographs). Bibliography: p. Includes index. 1. Plants, Protection of. 2. Plant parasites – Control. 3. Micro-organisms, Phytopathogenic – Control. 4. Agricultural extension work. I. Palti, J. II. Ausher, R., 1936–. II. Arenstein, Z. IV. Series. SB950.A36 1986 632 85-23241.

Typesetting, printing, and binding: Konrad Triltsch, Graphischer Betrieb, Würzburg
2131/3130-543210

Preface

Advisory work, by its very nature, is an intermediary between the research worker and those who apply the results of his research. The challenge of advisory work is to devise means of and find pathways for transmitting research results to the user, overcome the reluctance of the latter to change, and often combine novel ideas with well-established traditions. Nowhere is this challenge greater than in farming. This is especially true in developing countries, where the gap in the educational level between research workers and farmers may be extremely wide. Moreover, village-level advisers are often overburdened with non-professional functions and are not sufficiently backed up by well-trained professional advisers. Thus, in many of these countries there is a serious discrepancy between the knowledge available and that needed and actually applied on the farm.

Advisory work in crop protection is no exception, but profits to some extent from two facts: (1) because of the potentially catastrophic nature of pest attack, governments often operate a supervisory crop protection service, the staff of which may be able to dispense some pest control advice; and (2) the staff of pesticide distributors tends to fill, at least in part, the need for advice on how to fight pests and diseases with chemicals.

In recent years improvements in crop protection advice in many countries have taken place, not least due to the emphasis placed on this work by international agencies. A concerted effort is being made to generate up-to-date knowhow and disseminate it efficiently.

In this book it has not been possible to cover the entire range of crop protection advice. Weed control has not been included, and important pests such as rodents and birds are not dealt with. Pests and diseases of forests are not covered. In order to keep the book within its limits, we have refrained from discussing general aspects of advisory work, such as mass communication, advice on decision-making in general, and have concentrated on aspects that have a bearing on crop protection advice.

One difficulty to be overcome when writing about agricultural advisory work in English, is that of terminology. In Britain, and in most countries in which Ministries of Agriculture have been set up on the British model, the staff instructing farmers are called advisers. But in most of the United States farm advice has first developed as a service extended to the farmer by local universities; it has therefore been termed extension work, and this term has also been adopted by many international agencies. In this book we have opted for the designation of the work described as "advisory work", except for the chapters contributed by colleagues from the U.S.A. We also prefer the spelling "advisers" to the American "advisors".

The term "specialist" also requires precise definition. We think it should best be used in a restricted sense. Thus advisers for the whole farm or for groups of crops are here referred to as crop advisers, and the term specialist is reserved for advisers whose work is limited to one or a few crops only, e.g., cotton specialists. Similarly, in crop protection work, the adviser covering many aspects will be referred to as crop protection adviser, and the adviser specialized by pest control discipline as advisory phytopathologist, entomologist etc.

The designation "subject matter specialist" is commonly used, especially in developing countries, in referring to advisers on certain crops or on more narrowly circumscribed crop protection problems. With their deepening knowledge on particular subjects, these specialists serve to back up field advisers who have to deal with a wide range of cropping problems. The term "subject matter specialist" has here been retained in describing advisory work in the developing economy of Thailand (Chap. 3.7).

Crop protection advisers are a decidedly heterogeneous group, ranging from academically trained staff, who keep up with current literature and have access to libraries and auxiliary services, to farmers with plenty of experience, but little formal training and limited opportunity to keep up with new developments. We have therefore endeavoured to take into consideration those of our readers who are not too familiar with recent literature and innovations. This has caused certain differences in the professional level at which some of the subjects are presented. Thus the chapters on diagnosis cannot give more than a rapid overview of this crucial aspect of crop protection advice, and the chapter on prognosis is limited to the review of some relevant examples. Somewhat more detailed descriptions are given of the concepts of thresholds and of sampling, as applied to crop protection, and the thorny subject of pesticide application is treated at some length. Emphasis has throughout been laid on practical application of the scientific principles involved.

The chapters describing advisory work in pest and disease management in various countries present examples of the scope and organization of this work, the concepts involved and their translation into practice. A full account of the present advisory set-up is provided for England and Wales, for the Federal Republic of Germany, for California, and for Israel. Other authors have focused on particular aspects, such as the organization of Integrated Pest Management in Texas, and the dissemination of advice in the Midwest of the U.S.A. Finally, the chapter on Thailand aims to show how the so-called

"Training and Visit" approach to extension work is being used to develop a crop protection advisory service in developing countries.

Latin names of pests and pathogens appear in the text without authors' names, but the latter are given in the Index of Pests, Pathogens and Beneficial Organisms at the end of the book.

The Editors

Acknowledgements

The editors wish to thank Prof. J. Kranz, Tropeninstitut der Justus-Liebig Universität, Gießen, for his helpful comments, and to the late Y. Teich for his advice on entomological aspects. Many colleagues in the Extension Service of the Israeli Ministry of Agriculture have given us the benefit of their expert advice, and among them we thank most particularly Mr. H. Shoham and Mr. S. Moeller. Dr. Q. Argaman, Department of Plant Protection, has been most helpful in checking the taxonomic designations of the pests mentioned.

Contents

Contributors

ARENSTEIN, Z., Department of Agriculture, Agan Chemical Manufacturers, P.O. Box 262, Ashdod, Israel

AUSHER, R., Extension Service, Ministry of Agriculture, P.O. Box 7054, Tel Aviv 61070, Israel

BEICHT, W., Landespflanzenschutzamt Rheinland-Pfalz, Essenheimer Straße 144, 6500 Mainz-Bretzenheim, FRG

BLAIR, B. D., JBC Consulting Incorporated, 10546 Tarlton Road, Circleville, OH 43113, USA

BLAZQUEZ, C. H., Agricultural Research and Education Center, Lake Alfred NASA Kennedy Space Center, Lake Alfred, FL 33850, USA

DINOOR, A., Faculty of Agriculture, Hebrew University, Rehovot, Israel

DISHON, I., Agricultural Research Organization, P.O. Box 6, Bet Dagan 50250, Israel

EDELBAUM, G., Division of Remote Sensing, Extension Service, Ministry of Agriculture, P.O. Box 7054, Tel Aviv 61070, Israel

ELKANA, Y., Extension Service, Ministry of Agriculture, P.O. Box 7054, Tel Aviv 61070, Israel

ESHEL, J., Chemical and Technical Supplies Ltd., P.O. Box 10, Tel Aviv 61000, Israel

FLETCHER, J. T., Agricultural Development and Advisory Service, Ministry of Agriculture, Fisheries and Food, Olantigh Road, Wey, Ashford, Kent TN25, 5E1, United Kingdom

FRANKEL, H., Institute for Farm Engineering, Agricultural Research Organization, P.O. Box 6, Bet Dagan 50250, Israel

FRISBIE, R.E., Integrated Pest Management Coordinator, 311 Systems Building, Texas A & M University, College Station, TX 77843, USA

GENIZI, A., Division of Statistics and Experimental Design, Agricultural Research Organization, P.O. Box 6, Bet Dagan 50250, Israel

GOLAN, J., Extension Service, Minstry of Agriculture, P.O. Box 7054, Tel Aviv 61070, Israel

HANUSS, K., Landespflanzenschutzamt Rheinland-Pfalz, Essenheimer Straße 144, 6500 Mainz-Bretzenheim, FRG

HOCHBERG, R., Division of Remote Sensing, Extension Service, Ministry of Agriculture, P.O. Box 7054, Tel Aviv 61070, Israel

McWHORTER, G. M., 411 Systems Building, Texas A & M University, College Station, TX 77843, USA

PALTI, J., Agricultural Research Organization, P.O. Box 6, Bet Dagan 50250, Israel

SACHS, Y., Extension Service, Ministry of Agriculture, Hadera, Israel

THOMASON, I. J., Department of Nematology, University of California, Riverside, CA 92521, USA

TOSCANO, N. C., Cooperative Extension, University of California, Riverside, CA 92521, USA

Part 1 Advisory Work in Pest and Disease Management in the General Framework of Agricultural Production

1.1 The Place of Pest and Disease Management in the Agricultural Economy, and Its Legal Framework

J. Palti and R. Ausher

1.1.1 Communal, Risk, and Emergency Aspects of Crop Pest and Disease Control

Pests and disease may hit one farmer's crops more severely than those of his neighbours, but their control is essentially a community problem, and many control aspects should be approached on a communal basis. One farmer taking the risk of an epidemic in his crop may endanger others in his region, and this conveys on crop protection a status not shared by any other discipline in crop production.

Two other aspects, those of the risk involved and the frequency of emergencies, derive from the fact that losses from pests and diseases are still to a marked extent unpredictable.

1.1.1.1 Risk

The emphasis in farm management is frequently not on obtaining maximum yields, but on evading avoidable risk in the field and in marketing. There is not much the farmer can do to avoid weather-related risk. However, the majority of important pests and pathogens are amenable to treatments that limit potential loss. The evolution of treatment thresholds for numerous pests and diseases in countries with well-developed farming economies has done much to tell farmers when crops are in danger and should be treated. But farmers are highly averse to risk, and those growing high-value crops cannot afford even slight yield reductions. There is thus a tendency for "insurance" treatments, often applied prophylactically. The introduction of systemic pesticides active even when applied with some delay promises some reduction in the number of such applications. But the insurance value of pest and disease control operations will and should remain one of the aspects prominent in the farmer's mind.

1.1.1.2 Emergency

Pest and disease control operations can be planned only up to a point. Influx of pests or pathogens from extra-regional or even quite distant sources may surprise the farmer. More frequently, his failure to recognize, or his neglect of, low initial infection or infestation levels may force him to perform emergency control operations of the "fire-brigade" type. Such emergencies demand from the grower a high degree of flexibility and a capacity to take decisions rather more quickly than those taken in most other aspects of farming.

1.1.2 The Legal Framework for Pest and Disease Control and for the Adviser's Work

1.1.2.1 Importance of the Legal Framework

The principal reasons why pest and disease control in crops is so often a matter of public interest can be listed under three headings:

a) Pest and diseases appearing in one farmer's field may well endanger the crops of his neighbours.
b) Noncrop areas under public management (waterways, roads, etc.) may be the source, or contribute to the persistence, of crop parasites; noncultivated areas frequently harbor rodents and other animal pests potentially injurious to crops.
c) The use of pesticides in one crop may affect other crops in the vicinity and may endanger fish, bees, wildlife and water sources.

All matters of public interest require a legal framework, and so does crop protection. Relevant laws, rules and regulations provide the framework for all advisory work in crop protection. This is important everywhere, but most particularly so where representatives of pesticide producers are active in advising farmers on the use of chemicals and where private consultants furnish advice.

1.1.2.2 Aspects to be Covered by Law

Regulations should cover all or most of the following aspects:

a) Supervision of propagating material, including seeds, tissue cultures, transplants, grafting material, etc., which may carry pests or diseases.
b) Licensing of pesticides, with due indication of the range and duration of their action, phytocidal properties, dosage, toxicity to man, dangers to wildlife, and period required to elapse between application and harvest in each crop destined for human or animal consumption.
c) Protection of bees and wildlife from pesticide damage.
d) Licensing of operators for use of highly toxic chemicals.
e) Licensing and supervision of application equipment used by commercial operators.
f) Instructions for medical treatment of operators if toxic chemicals are used.

g) Regulations for the sale of pesticides and for their storage, with special reference to their deterioration in storage or after containers have been opened.
h) Definition of minimum distances between areas treated with a given chemical and the fields or plantations to which such pesticides may drift and where they may leave undesirable residues, cause phytocidal damage, or affect beneficial insects or mites.

1.1.2.3 Legal Aspects and the Adviser's Work

The above regulations in many ways involve duties and impose constraints on the work of the crop protection adviser. Though the latter should not be a law-enforcement officer (see Sect. 2.6.1.2), he cannot countenance transgressions and must advise farmers and crop advisers how best to manage within the limits of the law. In the following, some duties of the adviser in relation to relevant legal prescriptions will be discussed:

a) *Propagating Material.* Supervision of the health of propagating material is of paramount importance. In view of the fact that such material can be the principal source of pathogens, and frequently also of pests, and considering the expert knowledge required to detect potential danger on seeds and vegetative propagating material, the crop protection adviser carries a definite professional responsibility in this field.
b) *Licensing Pesticides.* The voice of the crop protection adviser should always be heard in the process of licensing pesticides. Advisers can contribute greatly to licensing committees, because they are best able to assess the practical needs for and consequences of releasing a pesticide for any particular use, or of denying or delaying its release. This is particularly true where new groups of pesticides are concerned. Unless health hazards are involved, the earlier or later licensing of such pesticides often depends on the urgency of problems they are meant to solve, and on the alternative available: nobody is as well-placed, or in many cases as competent, as the crop protection adviser to consider alternatives in such situations.
c) *Training Pesticide Applicators.* It often is the crop protection adviser's task to impart to pesticide applicators the necessary knowledge on (1) risks of working with toxic chemicals; (2) conditions that affect the efficacy of pesticides, such as wind, leaf wetness, imminent rain, and (3) hazards to neighbouring fields, especially in working with herbicides.
d) *Supervision of Application Equipment.* Advice concerning proper maintenance and use of pesticide application equipment is of general importance, and particularly so where growers have their crops sprayed by commercial operators. Among the aspects to be considered are the uniformity of coverage (see Chap. 2.5), use of mixed (combined) sprays, wheel damage and risks to neighbouring fields (see below).
e) *Emergency Treatment of Pesticide Operators.* The plant protection adviser should be aware where emergency instructions are available for the treat-

ment of pesticide applicators working with toxic pesticides, and where antidotes are within easy access either on the farm or in the village.

f) *Storage of Pesticides* is of importance because (1) they must be kept out of reach of children and unauthorized persons; (2) they may deteriorate rapidly if not stored properly; and (3) even in proper storage, many pesticides gradually lose potency: unless the farmer is suitably instructed, he may use pesticides over longer periods at inadequate rates, and suffer losses.

g) *Risk to Neighbouring Crops.* Effect of pesticides on crops bordering the crop treated, or even over longer distances, obviously depend on wind conditions and droplet size, as well as on the selectivity of the material used. Farmers require pertinent advice regarding the mode of application (including nozzles, pressure, and everything else that affects droplet size) to minimize the risk of drift (see Chap. 2.5). This is most important with herbicides and with broad spectrum insecticides that may affect beneficial insects in crops in the vicinity.

To conclude this review of legal aspects, it should again be stressed that no crop protection advice whatsoever must be given outside the legal framework. Crop protection advisers can play an important role in training, on the one hand, public and private advisers in observing relevant laws, and on the other hand, in acquainting law-enforcement officers with the problems posed by some of the legal constraints. However, as mentioned elsewhere, law-enforcing duties should not be imposed on the adviser.

1.2 Crop Protection Advice: Its Place in the General Scheme of Farming and of Farm Advice

Y. ELKANA, J. PALTI, and R. AUSHER

1.2.1 Overall Phytosanitary Management – the Basic Aim of Crop Protection Advice

Crop protection, in its early stages, has developed as a number of separate scientific disciplines, including entomology, phytopathology, nematology, weed science, etc., which proceeded on parallel paths rather than in unison. In the case of entomology, in particular, this was largely due to the preponderance of medical entomological research. Weed science was just a part of botanical research until close to the 1950's.

Modern crop protection, in its research, extension and application, though still often given along specialized lines, tends to proceed in the direction of a unified approach to all that ails the crop. This is due to many factors:

a) The intensification of agriculture and advanced cultural techniques; e.g., if you change the irrigation regime or crop sequence in a crop, you also change the habitat for all pest and pathogen development (Palti 1981).
b) The pronounced importance assumed by the applied side of crop protection, in which advisory personnel plays the essential part of integrating cultural practices and specific crop protection measures.
c) Recognition of the nature of nonparasitic disorders and of their interactive relationships with many parasites.
d) Biological understanding, especially recognition of the fact that a treatment controlling one harmful organism may promote another.
e) Resistance breeding, since there is no sense in breeding a crop resistant to one disease, if it is then liable to be attacked by another destructive pest or disease.
f) Economic considerations, especially establishment of treatment thresholds for combined effects of major pests or diseases that may affect a crop. The higher yields obtainable by modern farming methods justify control treatments where these were not economic at lower levels of yield.
g) The overwhelming role of the chemical industry in promoting the use of pesticides for the protection of crops.
h) The pronounced progress in the use of mixtures of insecticides, miticides, fungicides, and sometimes herbicides, for combined control of several parasites.

i) The spectacular widening of the use of herbicides, and the vistas opened up by such weed control in updating traditional practices, such as crop rotation, which affect a wide spectrum of pests and diseases.
j) Development of regional concepts of crop protection (Palti 1981).
k) Better understanding of the impact of cultural and crop protection practices on post-harvest pathogens.

The aim of to-day's crop protection efforts should therefore be defined as *overall phytosanitary management*, which relates to everything that may harm the growth of the crop and lower the farmer's revenue. This goes beyond the systems for integrated pest management for individual crops, as introduced in various countries and especially in the United States (see Chap. 3.6).

Overall phytosanitary management includes a regional, inter-crop approach with equal attention to crop-specific problems and to those posed by multi-host parasites, whether they be moths, mites, sucking pests that transmit virus diseases, or air-borne pathogens with wide host ranges, e.g., *Sclerotinia* fungi, dodder, and others. It also includes control of organisms attacking plants both in fields and on non-agricultural land (rodents, weeds) (see Sect. 2.4.7.4).

1.2.2 A Crop Worth Protecting Is the Prerequisite for Crop Protection

Crop protection advice is of value only where a crop worth protecting is at risk. The importance of crop protection in a farm economy thus rises according to the risk inherent in growing the crop, apart from the risks stemming from force majeure. Crop protection is not an important factor in low-level, subsistence farming where production proceeds in a more or less stable agro-ecosystem, and epidemic disease or pest outbreaks are rare (Putter 1981). If such systems are upset, the problem is less one of advice than the need for urgent implementation of treatments to avoid famine.

The higher the yield and income the farmer aims at, the higher the risks he has to take in his practices, and the greater the need for crop protection advice. But only when the crop has been suitably planned as regards season, soil, density of stand, irrigation, etc., i.e., only when its cultural background is properly established, can crop protection advice bring full benefits.

Ideally, crop protection advice should be given to farmers side by side with general cropping advice. This is best done, for all but highly advanced growers, by the crop adviser who, in his turn, is advised and kept up-to-date by the crop protection adviser and research staff.

However, in practice, in both developed and developing countries, wherever crop advisers do not reach growers with adequate frequency, crop protection advice is often given without relation to general farm advice. This has two principal reasons: (1) Some developing countries, which do not yet have a general farm advice system, have put up a crop protection service to deal with legal considerations and with the "fire-brigade" aspects of crop protection work; this rightly includes some crop protection advice. (2) Pesticide manufacturers and distributors have large numbers of advisory and sales personnel in the field, and these quite often are the advisers that visit the farmer most frequently (see Chap. 2.6).

Thus crop protection advice, especially on the use of pesticides to control pests and diseases, is often given in a relative vacuum of crop-growing advice. This can be effective as ad hoc advice to stop epidemics, but does not contribute much to the evolution of balanced crop protection practices.

1.2.3 Aspects Particular to Advisory Work in Pest and Disease Control

In two principal respects, advisory work in pest and disease control differs essentially from the advice on cultural practices dispensed by the general crop adviser. These are the breadth of the scope and the measure of urgency inherent in crop protection.

1.2.3.1 Multi-Crop and Multi-Annual Aspects

The crop adviser is concerned with a certain group of crops, which may to some extent influence succeeding crops, but do not importantly influence, nor are they influenced by, crops in their neighbourhood, except for possible parasite transmission. For the crop protection adviser the potential for such transmission is a central issue: he has to keep informed on the health situation of all crops in his region, to spot dangers of parasite invasions from field to field and from field to orchard and vice versa, and also, from non-agricultural land and roadsides to crops. Dangers of such parasite transmission may extend over several seasons, if pests or pathogen inoculum can overwinter or oversummer in the soil, in crop debris or on volunteer plants, and then attack crops in the same or neighbouring fields or orchards.

1.2.3.2 Urgency of Advice

Agricultural advice in general consists largely of planning crops, followed by current recommendations, mostly in reponse to changes in weather, as the crop develops. This is largely scheduled work, which is not too difficult to plan. The routine element is certainly not missing in pest and disease control work, but it is supplemented by elements of emergency: sudden outbreaks of pests and diseases require immediate, urgent attention, and competent advice can mostly be given only after ad hoc visits to the field or orchard affected. In the planning of advisory work for pest and disease control, a suitable, and sometimes quite considerable, reserve of time must be allotted to the unexpected, and this entails some difficulty in proper programming of such work.

1.2.4 Integration of General Cropping and Pest Control Advice

General cropping and pest control advice must be properly integrated. How this is done, and especially how much of the advice to farmers is given by crop advisers and how much by specialized crop protection personnel, depends on a few factors, some of which are mentioned here as well as in the section on the crop protection adviser and his tasks (Chap. 2.6). The factors to be referred to here are:

a) The risk of loss from pest and disease involved in growing prevalent crops.
b) The advisory personnel available, their proficiency, and the frequency with which crops can be visited, taking due note of commercial advisers who may visit the farmer.
c) The educational level of the farmer.
d) The availability of auxiliary services, especially plant clinics, and of mass media for the dissemination of warnings and advice.

1.2.4.1 Risk of Crop Loss

Where the risk of crop loss from pest and disease is low, the crop adviser can supply most of the advice required. Where medium- or high-risk crops are grown by farmers of limited vocational education, they require frequent inspection, which is in general provided by scouts, sometimes by crop advisers, rarely by crop protection advisers. In such crops, much of the crop adviser's work is concerned with crop protection, constitutes much of the "message" he brings the farmer, and enhances his status. But in these crops, close contact between the crop adviser and the crop protection adviser is of the utmost importance. Many growers of high-risk crops, such as greenhouse or flower crops, are at a high professional level, familiar with desirable cultural practices, and require current advice chiefly on pest and disease control; such growers are often best served directly by crop protection advisers.

1.2.4.2 Availability and Proficiency of Advisory Personnel

The number of advisers available, their professional proficiency, and the number of farms each has to care for, are key factors in general as well as in crop protection advice. A distinction must here be made between farmers who will not or cannot, at their own initiative, ask questions and invite visits, as is the case in many developing countries, and farmers who are capable of recognizing a problem when they see one, and call in the adviser or telephone for advice. In the former case, advice must be given by routine visits, which have to be made by crop advisers, whatever their number or proficiency. In the latter case, much more of the advice for pest and disease control can be given by crop protection advisers, working closely with the crop adviser responsible for the area or crop concerned.

1.2.4.3 Farmers' Educational Level

Where farmers are uneducated, at the local village level, in developing countries, and where they do not readily adopt new practices, all advice has to be united in the hands of the general crop adviser. There is no room for specialized advice in pest and disease control to the farmer, but the crop adviser should currently be advised and brought up-to-date by crop protection specialists.

As the farmers' professional level and adaptability to changing conditions (weather, market) improves, the role of the crop protection adviser in working directly with the farmer, becomes gradually more important. But, for all but the best-educated farmers, the crop adviser is still the principal source of advice.

1.2.4.4 Availability of Auxiliary Services

Among these services, those provided by the plant clinic are of the greatest importance (Chap. 2.2.3). Where such clinics are available, staffed competently, and provide diagnosis with reasonable speed, their staff and data bases play an important role in replying to questions of educated farmers, but more particularly to those of crop advisers and crop protection advisers; they are valuable in developing the advisers' ability to cope with problems posed by parasites not easily diagnosed in the field, especially soil-borne pathogens and viruses. In fact, in some countries such clinics have become a well-integrated part of the advisory services (see Chap. 3.1).

Other auxiliary services to be mentioned here are the mass media. Since dissemination of advice in pest and disease control is often urgent, and frequently affects wider regions, access to radio, television, and newspapers enables such advice to be dispensed to farmers and crop advisers at short notice.

Both plant clinics and mass media, if used to proper purpose, may widen the area that can be served by individual crop protection advisers, and make the field work of the crop adviser in the sphere of pest and disease control more effective.

1.2.5 The Target Groups for Advisory Work in Pest and Disease Control

As mentioned above, straightforward and current pest control advice should in general be given to farmers by crop advisers rather than by crop protection advisers. The task of the latter should more commonly be to devise pest control strategies, and to place suitable options before those who can weigh them competently and take their own decisions. This is of particular importance in the development of integrated pest management. The target groups of the crop protection adviser are therefore often advanced farmers, managers of large farms, plantations, or agricultural cooperatives, and in many cases not so much the farmers themselves but crop advisers, inspectors, and field scouts. These groups may be expected to be better-trained, and sometimes also more experienced, than the majority of farmers. This distinction is less apparent in developed than in developing countries; in some of the latter, especially in Africa, numerous local dialects spoken by the farmers in any case may make direct crop protection advice by anyone coming from another district most difficult, but village chiefs and local crop advisers may be expected to speak both the local dialect and a language common to larger regions or whole countries. These would then form the principal target group for the crop protection adviser.

Agricultural schools, at almost any level, are to be included in advisory work in crop protection. Preparation of teaching aids and leaflets for schools may devolve on the crop protection adviser and be a worth-while investment to promote pest and disease awareness in farming communities.

The field staff of pesticide manufacturers, distributors, and applicators is an important target group for crop protection advice and for the exchange of views.

Additional target groups include agricultural planning staff, crop insurance personnel, valuers of yields and yield losses, and private crop consultants.

1.2.6 Organizational Concepts for Integrating Cropping and Crop Protection Advice

In order to meet the requirements of crop protection advice under conditions of extensive as well as intensive, developing as well as developed agriculture, five general concepts and organizational forms have evolved for its integration with over-all cropping advice. There also exist many intermediate forms adapted to local requirements. These five concepts are:

a) All advisers at the farm community levels are general crop advisers, who often include animal husbandry as well as crops in their work. Crop protection specialists serve at district, region, or state level only. This applies to the situation in Kenya, India, Peru, and some states of the U.S.A., to cite just a few examples.
b) Both crop advisers and crop protection advisers serve at the local level, within the framework of general farm advice. Some of the crop protection advisers may be specialized by groups of crops or by disciplines. They are supported by higher level experts working at regional centres or research institutions (most west European countries, Israel, and the U.S.A.).
c) Crop protection advisers operate outside the structure of general farm advice, either in separate units of the Ministry of Agriculture (e.g., in Thailand, Togo and the Federal Republic of Germany) or in research institutions (e.g., Portugal).
d) Where agricultural production is predominantly organized in large farms, the ministry provides overall guidance, sometimes including warnings, but the farm employs its own crop protection adviser. This is the advisory sys-

tem found in some east European countries, on cooperative farms run by the state, but also in developing countries in which crops are primarily produced on large plantations or in coordinated regional projects.

e) The advisory organization has no crop protection advisers of its own, and depends chiefly on the advice given by commercial firms. With the improvement of advice provided by such firms, this is frequently found to be the case in some regions of southern and western Europe.

1.2.7 A Brief Look at the Range and Limitations of Advice by Public Agencies and by Pesticide Distributors

Where the use of pesticides is concerned, all crop protection advice, no matter by whom it is given, must conform to the prescriptions of the pesticide registration (licensing) authorities. This includes prescriptions on (1) the concentrations at which the pesticide has been found effective against the target pathogens or pests, without causing phytotoxic injury; (2) the period that must elapse between application and entry of personnel into the field and, most importantly, before the produce is harvested, and (3) any other conditions or constraints that affect the efficacy and safety of application, e.g., extreme weather conditions to be avoided, potential problems of cumulative deposits after more than one application, etc.

These prescriptions, as indicated in general on the label, constitute the framework of advice on crop protection by chemicals. But whithin this framework there is often a great deal of room for manoeuvering. Where several pesticides of equal efficacy and safety can be chosen from, the adviser on the staff of public agencies may be expected to choose the cheapest and most readily available material, or perhaps the one the farmer has in store, taking into account health hazards to operators, compatibility with other pesticides, application equipment at hand, and weather and crop conditions.

The adviser on the staff of pesticide distributors unavoidably works under commercial constraints. Other things being equal, he will recommend the product his company wishes to sell; he may be better informed than anybody else on two points of great interest to the farmer: the price and availability of the pesticide, specialized modes of application (e.g., fumigation) and compatibility with other pesticides that his company markets.

The field staff of pesticide distributors is in many countries more numerous, and visits farmers more frequently, than advisers in the services of public agencies. It is most important that the two groups share ideas on the choice and application of pesticides, and on the cultural and legal limitations to such applications. Constructive exchange and coordination of views of advisers in the two groups may avoid a great deal of confusion on the part of the farmer (see Chap. 2.6).

This is not difficult to achieve where advisers of a single pesticide distributor, with a wide range of pesticides, provide the advice. However, where representatives of several companies concurrently visit the farmer, conflicting advice may sometimes be given and confuse the farmer. In this situation the public crop protection adviser should give the lead in formulating advice.

References

Palti J (1981) Cultural practices and infectious crop diseases. Springer, Berlin Heidelberg New York, 243 pp

Putter CAJ (1981) The management of epidemic levels of endemic diseases under tropical subsistence farming conditions. In: Palti J, Kranz J (eds) Comparative epidemiology. PUDOC, Wageningen, pp 93–103

1.3 Basic Knowledge and Current Information for the Crop Protection Adviser

J. PALTI

1.3.1 Basic Knowledge

That the adviser for pest and disease control requires thorough training in all the disciplines of crop protection, including not only the pests and diseases, but also the techniques for controlling them, goes without saying. No less important is sound knowledge of general husbandry and of basic economic concepts, and this will be discussed here.

1.3.1.1 The Crop in Health

The crop protection adviser must be thoroughly familiar with all aspects and all stages of normal, healthy crop growth. This involves, inter alia, close observation of the development of new crops and varieties under prevailing local conditions and with standard growing practices. Since farmers often vary in their sowing, fertilization, irrigation, cultivation, and harvesting practices, they afford the adviser opportunities for observing the effect of such practices, under varying weather conditions, on each stage of crop growth. Such observations are essential for crop protection work, no less important than specific pest and disease control aspects. In fact, one depends on the other: only knowledge of the crop in health makes it possible to spot deviations from normal growth, i.e., potential disorders and maladies.

One aspect of crop growth particularly relevant to pest and disease control is the competent appraisal of the crop's potential for recovery and compensation, when harmed by pests, diseases, or non-parasitic factors at any one stage of development. This will be referred to in Chapter 2.2 on detection and diagnosis, and in Chapter 2.3 on thresholds for treatments against pests and diseases.

1.3.1.2 Innate Destructiveness of Pests and Pathogens

The damage that parasites can be expected to cause often bear little relation to their numbers or the area covered by their symptoms. This is obvious in two cases: (1) where insects serve as vectors for virus diseases and even few vectors may cause much infection; and (2) where marketable produce is directly affected, and even a single spot may make the fruit or flower unfit for marketing or reduce it in grade. However, apart from these cases, many parasites cause damage in no way commensurate with their numbers. This applies to organisms causing leaf, flower, or fruit drop, although infection may only be slight, to pests or diseases causing stem breakage, to parasites which may invade the plant at only one point but produce toxins that spread to affect wider parts of the plant, and to many other pests and pathogens that cause indirect damage. The grower, in attempting to appraise the extent to which his crop has been affected and in making up his mind whether or not to take control measures, may not be aware of the damage potential parasites not particularly abundant, and may need special instruction on this point.

1.3.2 Current Information for the Adviser

Crop protection work relates to many aspects of human endeavour, including economics and health, and is, of course, closely linked with the weather. Information currently available on such aspects must be sought by the crop protection adviser, and this will briefly be discussed here.

1.3.2.1 Potential Crop Value and Current Prices of Produce

The definition of treatment (or control) thresholds, a subject to be described at length in Chapter 2.3, depends partly on the price level at the farm gate of the produce to be marketed, stored, or processed at specific seasons, and on expenses directly involved in marketing, and partly on the prices of pesticides, labour and application equipment. All these prices are subject to change in free market economics and large growers are well aware of this. However, small growers, unless specialized and trained, may require expert advice in weighing the potential profit in the use of pesticides, and the crop protection adviser must constantly keep informed to proffer such advice.

1.3.2.2 Pesticide Registration and Compatibility

In countries with well-organized crop protection services, the registration of pesticides is a continuous process, and registration for use by growers is sometimes subject to important restrictions, such as limitation of period between the pesticide's last application and the picking of fruit, and potential danger of the pesticide to bees (see Sect. 2.4.7.3). It is, of course, of supreme importance that the crop protection adviser be aware of all newly registered materials, their properties and limitations. Relevant information is currently available in the label specification of registered compounds. In order to facilitate rapid and more sophisticated retrieval of such data, lists are now being made available by the licensing authorities in various countries by use of computerized data banks.

The subject of pesticide compatibility, i.e., which pesticides can be combined to be used in a single application, is one of the most delicate in all the field of crop protection by chemicals. Mixing tables are available, but often, when published by public bodies, only state which toxic ingredients are in principle compatible, do not react chemically with each other or do not otherwise change each other's pesticidal or phytotoxic effects. This, however, does not go very far: the formulations in which such ingredients are marketed may greatly affect their compatibility. Pesticide manufacturers publish mixing tables (compatibility charts) only for their own range of products, as they may not be fully aware of the nature of the formulation marketed by others, and understandably wish to avoid all responsibility for effects of sprays involving pesticides other than their own.

However, in day-to-day crop protection practice, the question whether or not two pesticides needed at the same time can be mixed is so important, that the question is constantly put to the adviser. Where mixing tables do not provide adequate answers, the adviser cannot assume the responsibility for recommending mixes not fully tested under varying external conditions. But he can gather practical field experience from growers who have tried the combinations in question. Such observation will be limited to the occurrence or non-occurrence of visible phytotoxic effects, but will rarely give a clear picture as to the pesticidal effect of the mix, as compared with the effect of each ingredient applied separately. A constant interchange of information on problems of compatibility should be an important part of periodic meetings of crop protectionists and up-to-date data bases should be available.

1.3.2.3 Health Risks and Antidotes

The application of substances toxic to humans is an integral part of crop protection practice. Accidents to operators, other field staff, and consumers should not happen where reasonable precautions are taken. However, "familiarity breeds contempt", and even seasoned operators are often negligent. Labourers are frequently unaware of the risks of handling toxic chemicals or of consuming in the field food that may have become contaminated.

In addition to constantly pointing out the dangers inherent in being exposed to toxic chemicals, there is a duty incumbent on the crop protection adviser: to know which antidotes can be used in case of accident, and to know where in his region these are available. He may even have to bring to the notice of health authorities that pesticides against which certain antidotes are effective are used in the region, to request stocking of such antidotes.

1.3.2.4 Weather Data

The weather plays so important a role in the health of the crop, that all of its aspects, whether past records or prognoses, must be of constant concern to the crop protection adviser. To begin with, the meteorological peculiarities of the region and its subregions should be defined as clearly as possible, including: dangers of frost, precipitation (dew and rain), winds and excessive insolation. For this purpose, full use should be made of the data bases available in most meteorological services. Then cooperation must be established with all stations that keep weather records, and with growers who do so. A constant flow of information on weather experienced or expected is the basis for the evaluation of risks of direct weather injury to crops, of the occurrence of stress conditions that may accentuate parasite trouble, and of conditions that favour parasites so much that outbreaks may be imminent. Computerized weather information services are now available in most developed countries. Such information may include specific crop protection components, such as calculation of degree-days for various pests (see Sect. 2.4.2.3), warnings of extreme weather conditions and indications of weather conditions (wind!) relevant to pesticide application (see Chap. 2.5).

1.4 Farmers' Perceptions of Pest and Disease Control

J. PALTI

The farmer's behaviour in matters of crop pests and diseases and their control is based on his perceptions of the problems involved, the attitudes dictated by tradition, and by his character (such as proneness to risk aversion or innate optimism or pessimism), the information he is given, the resources at his disposal, and the availability of means of control.

There is not much the adviser can do about the farmer's character. However, once he has gained the farmer's trust, he can certainly try to change certain perceptions, perhaps even traditions, and the way in which resources are employed for crop protection.

The subject of farmers' perceptions in relation to pest and disease management has only recently received the attention it deserves, especially in England (Mumford 1981, 1982, 1983, Mumford and Norton 1984, Norton 1976, Tait 1978). Some of the findings of this research will here be referred to in the context of pest management advice.

The farmer's perceptions relating to pest management are based primarily on:

a) His training and experience in diagnosing the causes of deviations from the normal in the development of his crop.
b) His capability of assessing realistically the loss hazard posed by crop parasites and the chances of crop recovery after it has suffered parasitic or non-parasitic injury.
c) His assessment of the effectiveness of the chemical and cultural control measures he can take with the time, labour, equipment and pesticides at his disposal (Norton 1976).

The adviser's work begins with improving the farmer's capability of diagnosis. One of the best ways of doing this is to take farmers on tours of numerous fields, pointing out the patterns in which non-parasitic disorders and pests and pathogens appear (see Chap. 2.2) and teaching recognition of salient symptoms by which parasites can be discerned. This is a straightforward matter of instruction.

Improving the accuracy of assessment of crop hazards and of the effectiveness of control measures is much more complicated. Such assessments are often not primarily a matter of judgement, but one of attitude to risk ("I'll be all right!") and of resources ("Can I afford the expense of treating or the risk of not treating the field?").

Mumford (1981), in a study of English sugarbeet growers, found that they tended to assess the risk of losses from virus yellow with a fair degree of accuracy, but to overestimate the efficacy of pesticidal treatments. Aversion to risk, and therefore use of control treatments as insurance, was common.

To dissuade anybody from taking out insurance, although it may appear superfluous, is a thankless task and one the adviser usually cannot risk. If the crop is "under-insured", i.e., not sprayed at all or not frequently enough, and then suffers losses, the adviser may be held responsible and may lose that all-important asset, the farmer's trust. On the other hand, the waste involved in overspraying is not perceived so clearly. What the adviser can do is to teach farmers, as their perceptions improve, the concepts of treatment thresholds to guide his future decisions. A demonstration of spraying or not spraying a small part of the farmer's field, when compared with that farmer's standard treatments, can yield convincing results, e.g., in showing up overestimates of pesticide efficiency.

With some farmers the use of pesticides has become something akin to a habit. Since he has succeeded with their use on one crop and against one pest, he falls into the habit of risk evasion by routine spraying. Here the adviser faces the task of getting the farmer to differentiate between the nature and benefits of various pesticidal treatments, e.g., between a one-time operation of weed control that may give sudden, spectacular results, and repetitive, more costly spraying for pest control with results that may become apparent only in time, and depend in part on cultural and weather factors. But breaking a habit by force of reason is a difficult undertaking, as millions of smokers know only too well.

The benefits derived from pest control by cultural practices are those the farmer is least likely to perceive and appreciate. These benefits are gradual, often cumulative, and do not attract attention. Farmers quite often are quick to adopt new, disease-resistant cultivars, but are much less inclined to change their practices of soil cultivation, sowing, irrigation etc. This is an area in which advisers can do much for the advance of crop protection, especially in the context of integrated pest management.

Finally, the adviser has to recognize that there are farmers who just do not wish to be advised and, consciously or unconsciously, resist any attempt to change their perceptions. On the other hand, there are farmers so keen on applying potentially effective pest control treatments, even before these have been adequately tested and officially approved, that they have to be dissuaded from rushing into operations which may give disappointing results.

Perceptions and considerations of peasant farmers in matters of pest and disease control are being referred to in Chapter 2.3.9.

References

Mumford JD (1981) A study of sugarbeet growers' pest control decisions. Ann appl Biol 97:243–252

Mumford JD (1982) Perceptions and losses from pests of arable crops by some farmers in England and New Zealand. Crop Protect 1:283–288

Mumford JD (1983) Perceptions of pests and controls in pest management, vol II. 10th Int Congr Plant Protect, Brighton, p 949

Mumford JD, Norton GA (1984) Economics of decision making in pest management. Annu Rev Entomol 29:157–174

Norton GA (1976) Analysis of decision making in crop protection. Agroecosystems 3:27–44

Tait EJ (1978) Factors affecting the usage of insecticides and fungicides on fruit and vegetable crops in Great Britain: Farmer-specific factors. J Environ Manage 6:143–151

Part 2 Advice for Pest and Disease Control: Tasks and Tools

2.1 The Range of Crop Protection Advice Is Extremely Wide

J. PALTI

A wide range of activities is expected from the adviser for pest and disease control, and he or she is constantly faced with ever-changing, sometimes critical situations that demand action.

To begin with: *diagnosis*. Perhaps no other aspect of pest and disease control requires more combined knowledge and experience than that of finding the answer to the question: what ails this crop? Grogan (1981) has put this well in saying that "in practice, a plant diagnostician must play the combined role of a detective and plant ecologist in the broadest sense: each disease problem requires a thorough search for clues, obvious or subtle, of abiotic and biotic factors and their interactions that are possibly involved . . ." We are here primarily concerned with field diagnosis, as practiced by the adviser when first confronted with fields or plantations that fail to develop normally or show obvious signs of suffering (Chaps. 2.2.1 and 2.2.2), but shall also refer to the help to be expected from plant clinics (Chap. 2.2.3).

Once the cause of the trouble has been diagnosed as a parasite, the question arises: how dangerous is the affliction and where does the *threshold* for economically valid control operations lie. This subject has received much attention in recent years, and in a number of European and North American countries computer models help advisers and growers to determine suitable thresholds (Chap. 2.3). A prerequisite to such determination is formation of a clear picture of the size of populations of pests or affected plants in the field under observation, and this can only be obtained by adequate and reliable *sampling* (Chap. 2.4.1).

A large part of the time spent by the crop protection adviser in the field is likely to be taken up by *pest monitoring* or scouting, or by teaching others to do so. This is time-consuming and therefore expensive work, but can be guided and restricted by suitable sampling and forecasting schemes, including the determination of the presence or absence of pests or inoculum by trapping and other means (Chap. 2.4). Here again, computerized data can be of great help in directing the monitoring to the periods in which they do most good. Remote sensing is being developed as an aid for rapidly monitoring crops on a large scale (Chap. 2.4.5).

Once the need for chemical control has been established, growers have to be advised how to perform such control operations with least expense and greatest efficacy. This is perhaps the aspect of crop protection on which least knowledge is available to the adviser, and the subject of pesticide application will therefore be dealt with at some length (Chap. 2.5).

In many aspects, crop protection is essentially a *regional problem* (Palti 1981), because infestations in one grower's field endanger those of his neighbours, and treatments applied to one field may affect the growth of adjoining fields or their populations of beneficial insects. This presents the adviser with problems of the control of pests as well as of the perceptions of farmers (Chaps. 1.4 and 2.6).

The *tasks* a crop protection adviser should fulfil, and others he should preferably not be asked to fulfil require definition: a discussion of the merits and drawbacks of the adviser's *specialisation* by crop protection disciplines, and his relations with research institutions is presented in Chapter 2.6.1.

References

Grogan RG (1981) The science and art of plant-disease diagnosis. Annu Rev Phytopathol 19:333–351

Palti J (1981) Cultural practices and infectious crop diseases. Springer, Berlin Heidelberg New York, 243 pp

2.2 Detection and Diagnosis of Crop Disorders and of Pest and Disease Occurrence

2.2.1 Nonparasitic Disorders in Field and Orchard Crops

Z. ARENSTEIN, J. GOLAN, and J. PALTI

In the detection and diagnosis of nonparasitic disorders, the distribution of symptoms is often the first step. This comprises answers to two questions: are the symptoms distributed at random or in discernible patterns (e.g., along rows)?, and no less important: are symptoms restricted to one or a few plant species or are they more widely distributed over many species.

2.2.1.1 Stress and Injury

In dealing with nonparasitic disorders that are due to environmental stress, distinction must be made between symptoms indicating effects of reversible (elastic) stress and those of irreversible (plastic) stress (Levitt 1972). The most common symptoms of reversible stress are changes in leaf color or curling of leaves (see below). Reversible stress, e.g., that due to water deficiency or to temperatures that are low but above the freezing point, turns into irreversible injury if and when the threshold tolerance for repair by the plant is exceeded (Grogan 1981). This may be the case when the stress factor intensifies, or if the stress is exerted over long periods. When the adviser is faced with stress situations in the field or orchard, correct and rapid diagnosis is therefore of greatest importance, so that remedial measures can be applied while the stress effect is still reversible.

2.2.1.2 Patterns of Symptom Distribution

Spatial Distribution (Fig. 1). The first step in differentiating between nonparasitic disorders and crop afflictions caused by pest or disease is the recognition of the spatial patterns of their appearance (Anonymous 1968): if symptoms are scattered at random, with initially only few branches or organs of single plants or small groups of plants affected, the cause is most probably a pest or pathogen. If, on the other hand, large parts of the crop or entire rows are affected all

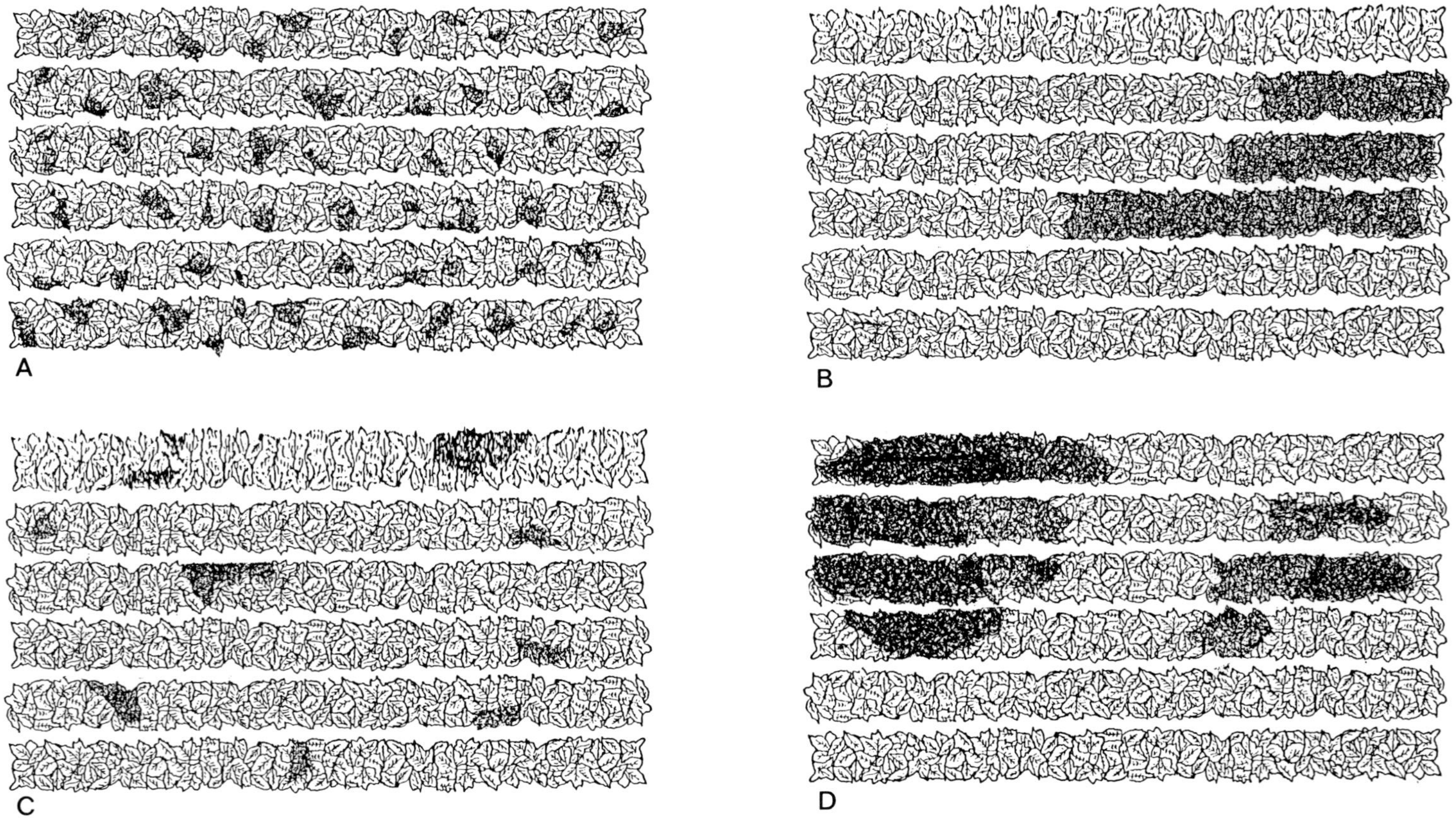

Fig. 1 A–D. Schematic representation of the spatial patterns of symptoms due to nonparasitic causes, pests or pathogens. **A** Generalized random distribution of symptoms, mostly due to weather factors. **B** Symptoms largely appearing along the rows or in more or less straight lines are generally due to man-made causes, e.g., scorching by chemicals. **C** Symptoms initially occurring as irregularly distributed foci may be suspected to be pests or pathogens, including fungi, bacteria, viruses, insects and mites. **D** Symptoms appearing in large patches are most frequently due to soil or water factors

at once, then the cause is to be sought in nonparasitic factors, whether environmental, such as hail, or man-made, such as scorching by chemicals (Fig. 2). But note: this distinction is valid only on two conditions: (1) that the propagating material has been free from pests or pathogens, since the latter, when borne on seed material or transplants are often distributed at random, and (2) that the crop is under continuous observation, and deviations from normal development are noted promptly. If this is not the case, the observer is unable to determine whether signs of malady have appeared suddenly and widely or not, and he may be too late to discover pests and diseases at the initial stages, when their appearance is still limited to individual plants or small groups.

The exception to the rule that symptoms due to nonparasitic causes appear simultaneously over large areas, is damage by lightning, which may cause large, isolated groups of plants to desiccate suddenly (Johnston and Booth 1983).

Symptoms due to disturbances in the water economy or nutrition of crops are mostly progressive, and eventually affect all or most parts of many field crops. Frequently, their first appearance is in medium-sized or fairly large parts of the field, e.g., where there are chalky patches in the soil or where water has accumulated in low-lying parts (Fig. 1 D).

However, in orchards, such symptoms may develop over several years, and may be apparent on leaves, fruits and young twigs before branches and trunks show signs of injury. An exception is injury by waterlogging which may cause sudden collapse of trees (see below).

In many crops, however, especially in fields and orchards that are not inspected very frequently, symptoms are sometimes observed on all or most of the older leaves, while they are not apparent on younger growth. This generally indicates injury due to a factor that has acted once, over a short time, but has then ceased to act. The most typical example is hail injury, which obviously affects only organs formed before the hailstorm, but not those formed thereafter. Other environmental factors, such as winds and rainstorms, may also be the cause of what may be called "one-time" injury, spread over the whole field. But if symptoms appear along the rows of row crops (Fig. 1 B), and perhaps on one side of the row or tree more often than on the other, "man-made" injury must be suspected: the cause may be top-dressings of fertilizers or applications of non-systemic pesticides, if their phytotoxicity has increased due to extreme weather conditions (e.g., hot, dry winds) or they are applied at excessively high concentrations. Such "one-time" symptoms contrast with those of many parasites which, unless controlled most efficiently, will currently proceed to affect growth as it forms.

Distribution over Plant Species. So far we have mentioned only spatial distribution of symptoms, but much can also be learned from their *"botanical"* distribution over one, a few, or numerous plants. If symptoms appear on a wide range of plants from different families, the cause may be assumed to be nonparasitic. If, for example, potatoes, a crop of the Solanaceae family, are affected together with weeds or neighbouring crops of other families, e.g., Gramineae or Leguminosae, this strengthens the presumption that the cause is nonparasitic. The same applies where tree crops as well as ground vegetation below them are affected. The exception may be where pests not selective in the choice of their food are concerned, such as locusts, and some birds or rodents.

Diagnosis of Air Pollution Symptoms. Since air pollution is most frequent in industrialized countries, where agricultural research facilities are readily available, diagnosis of symptoms of air pollution has received a great deal of attention (e.g., Mansfield 1983). Hicks (1976) has published a useful guide to the diagnosis of vegetation injury by pollution. Such diagnosis naturally begins with an attempt to identify potential sources of air pollution near the injured field, and continues with the consideration of terrain and location, e.g., valleys or basins in which air pollutants may be trapped. Further steps in diagnosis then concentrate on the questions formulated above, e.g., the distribution pattern of injury and the range of botanical species affected.

2.2.1.3 Recognition of Some Common Symptoms of Nonparasitic Disorders in Field Crops

Symptoms to be noted in the diagnosis of nonparasitic disorders in field and vegetable crops consist chiefly of changes in the colour or shape of the organs of the shoot. Some of the major symptoms and their most frequent causes are presented in Fig. 2, and are briefly discussed here.

Leaves Change Colour

- the most common causes for *yellowing* of leaves are excess water in the soil (waterlogging), excess of lime in the soil, and deficiencies of iron, zinc, or other minor elements (Sprague 1964).
- colours ranging from *purplish* to *pink* are primarily induced by low (but not freezing) temperatures, e.g., in Brassicae, clover, tomato.
- sudden *blackening* and *drying up* of leaves is an indicator of frost damage.
- *silvery* colour on the underside of the leaf is mostly due to sun scorch, which causes the epidermis to lift off underlying tissues.
- *whitening* of the entire leaf, or *vein clearing* or *interveinal yellowing* may be symptoms of herbicide injury.

Dry Spots on Leaves or Fruits

- *bright, sunken spots* on all parts of the plant, but especially on upper leaves and exposed fruit are most commonly due to hail.
- *larger dry spots*, especially on fruit, sometimes more frequent on that side of the crop which is exposed to radiation for longer daily periods, are apt to be due to sunscorch.
- *small dry spots*, often not randomly distributed over the field, but more severe in some rows (or parts of rows) than in others, may be caused by fertilizer scorch, or pesticide injury.

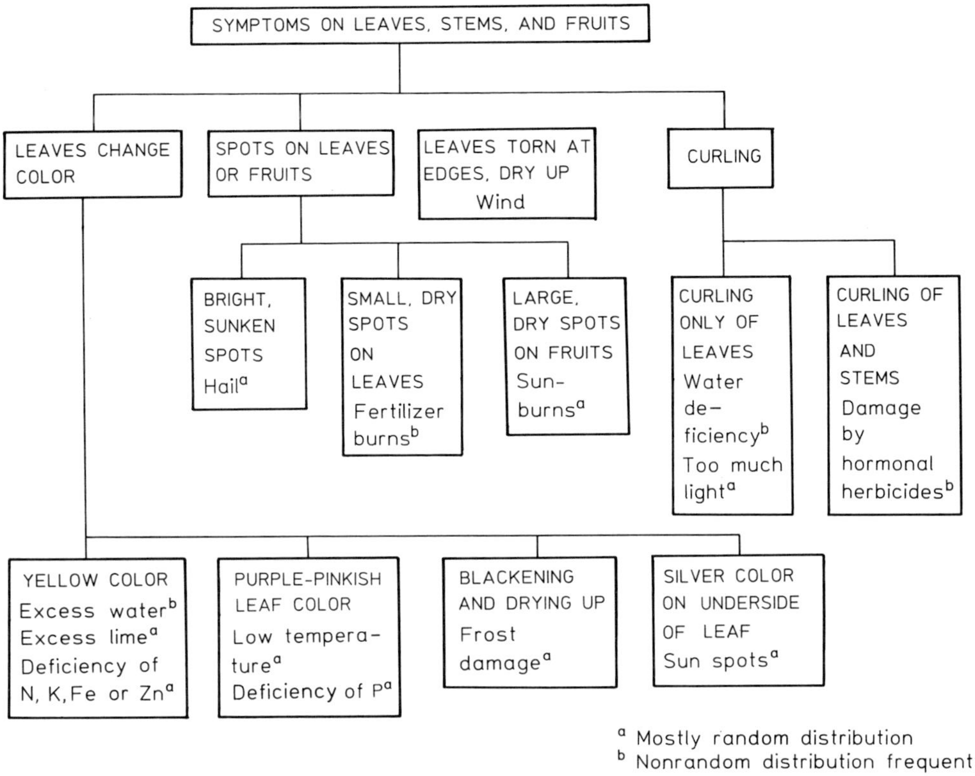

Fig. 2. Schematic representation of symptoms useful for the recognition of causes of some nonparasitic disorders in field crops. [a] Mostly random distribution; [b] nonrandom distribution frequent. (After Ausher and Shoham 1977)

Leaf Edges Torn and/or Drying Up

– these are effects of wind, and in dense fields or plantations are more frequent and more severe in leaves of plants at the edge of the field or otherwise exposed, than on plants sheltered to some extent, e.g., near windrows, buildings, etc.

Curling

– where the leaves only curl upwards or downwards, there may be two principal causes: if the leaves simultaneously show darker than normal green colour, the curling is apt to be due to water deficiency; but if there is no change in colour, and some of the leaves curling upwards show silvery colour at their edges, then excessively strong radiation may be the cause.

– where curling of leaves is accompanied by some twisting of stalks, this may be due to damage from hormonal weed-killers. Such damage is most frequently found in those parts of the field or plantation nearest fields sprayed with such weedkillers.

2.2.1.4 Recognition of Some Nonparasitic Disorders in Fruit Crops

Although, for the purpose of diagnosis, it is most convenient to deduce the cause of nonparasitic disorders from their symptoms, as has been described above for field crops, this has not been found practicable for fruit crops. The principal reason is the great diversity of these crops, which comprise (1) deciduous and evergreen, and low (1 – 2 m) to very high trees, bearing crowns of widely varying density, (2) wide ecological differences between young, widely spaced foliage and stems of young, as compared with those of older trees, and (3) the fact that most fruit trees are grafted, and rootstocks can exert great influence on the susceptibility of trees. To give just one example, the Nemaguard rootstock, widely used for peaches in Israel, affords the trees good protection against root-knot nematodes, but renders them highly susceptible to iron deficiency.

In view of this great diversity of fruit crops, and notwithstanding the fact that all that has been said in the previous section on patterns of distribution of non-parasitic disorders applies also to these crops, it has been found more convenient to describe disorders on fruit crops under the heading of the factors causing them: chiefly disorders caused by water factors, nutrient factors and stock-scion incompatibility.

a) Disorders Due to Weather Factors. Disorders due to extreme weather conditions occur on fruit trees in all climatic regions, but are perhaps more common in subtropical regions, where the culture of fruit crops originating in temperate and warm regions overlaps. There subtropical fruit suffers severely from occasional frost, while deciduous fruit suffers from excessively high temperature and radiation.

The most severe environmental injury, often leading to total loss of crops, is that caused by *frost*. Among subtropical fruits, avocado and banana trees hit by frost show total blackening of foliage and fruit; on citrus trees, among which lemons are the most susceptible, the leaves dry and woody parts crack and break, while the fruit exhibits no outward colour change, but its flesh dries.

Deciduous fruit and vine are obviously better adapted to freezing temperatures, and here the severity of the frost largely determines the symptoms: slight frost will cause blossoms to dry up and drop, leaves also dry up, the xylem in young twigs shows browning, and fruits show various symptoms, such as dark rings on apple fruits, but bigger branches and trunks will not suffer. However, when temperatures fall to -5 °C or below, dark symptoms of scorching appear along the bark of the trunk near the soil surface, the bark dies and cracks, and cankers are formed; in vines, all aboveground parts may die off.

Hail damage consists of sunken spots, holes, and tears on leaves, blossoms, fruits, and young twigs, with spots limited to the upper part of the organs affected. The outer, exposed parts of the tree suffer most, and blossoms may be destroyed completely. All tree crops may suffer if hail is severe. As mentioned above, formation of similar symptoms on the trees and on vegetation on the ground is a clue to hail as probable cause of the damage.

The injury caused by *winds* depends on their velocity, and on their temperature and humidity content. Strong winds injure leaves directly by tearing the tissue, especially of large blades (bananas) and indirectly by rubbing leaves and fruits against each other and against the woody parts of the tree: brownish lesions are then observed on leaves and brown-to-blackish ones on fruits, a frequent occurrence on subtropical fruit and citrus. If winds reach high velocities, fruits and leaves may drop, and branches, especially those weighted by fruit, may break off.

The damage caused by hot, dry winds consists chiefly in the scorching of blossoms, so that they wither and drop off, in similar scorching of young foliage, especially at the top of trees, and in the russeting of the peel of fruits of deciduous and subtropical crops. Irregular, necrotic spots often appear on citrus and avocado fruits and are due to mesophyll collapse.

Excessively High Atmospheric Humidity. When prevailing over long periods, this is often the cause of fruit russeting of apples and pears, and also of loquats. The tissue of affected fruits may crack.

High Temperatures and Radiation. High temperatures, usually above 30 °C, and concurrent strong radiation, or drastic fluctuations between day and night temperatures, affect in particular the fruit of deciduous trees, such as apples and plums, which show brown spots on their peel. Excessive radiation causes brownish rings around the lenticels of leaves of some deciduous and citrus crops: reddish discoloration of the tissue around lenticels sometimes produces rosy "cheeks" on the fruit of apricots and of certain apple varieties, and browning of grape berries. In the northern hemisphere, these phenomena, as well as sun scorch of exposed limbs or trunks, appear on deciduous and subtropical fruit trees most conspicuously on their southern or south-western side.

b) Nutrient and Water-Related Disorders. All such disorders cause first of all changes in leaf colour.

Deficiencies in *nitrogen, potassium,* or *phosphorus* cause leaves to assume light green, yellowish or reddish-brown colours, with necroses developing in general from the leaf margins inwards. In young, vigorous trees, nitrogen deficiency sometimes finds expression in light green leaf colour, tending to yellow, not only in the young foliage but all over the tree.

Deficiencies in *rare elements* usually take the form of partial or total yellowing of the foliage (Sprague 1964). Only a few examples can be given here. *Iron* deficiency results first in yellowing between the veins of young growth, but in an advanced stage of such deficiency the veins also yellow, and subsequently leaves whiten, become necrotic, and shed. *Zinc* deficiency produces various symptoms: on citrus trees it causes interveinal yellowing (mottle leaf), on apples small, dense leaves along the terminal twigs with lower parts of twigs partly or wholly devoid of leaves, and on pecan trees rosettes of terminal leaves are formed. *Manganese* deficiencies cause slight yellowing, extending from leaf margins toward the central vein, on citrus, apple, walnut and other trees. Excessive amounts of *boron* are highly toxic to some trees, including citrus, causing interveinal yellowing and shedding of leaves.

High Salinity. This at first causes the drying up of leaf margins, later of the whole leaf. Symptom expression may greatly depend on the rootstock. Avocadoes may show very pronounced symptoms of damage from high salinity, especially when grafted on rootstocks of Mexican origin, while citrus, apricots, and apples may also suffer, but not so strongly. Overhead irrigation with saline water can cause severe scorching of leaves, and even leaf-drop.

Water-Related Disorders. Lack of water causes leaves to yellow and dry up, and produces on the bark dark brown spots and cracks with pockets of resin.

Excess water, mostly due to waterlogging of heavy or badly drained soil, causes root rot and this finds expression in die-back of branches and, in severe cases, in the death of the tree.

Trees that are supplied with water after a prolonged period of drought, may sprout and blossom out-of-season; this may weaken the tree, as observed in particular in deciduous fruit crops.

c) Incompatibility of Rootstock and Scion. The appearance of bottle-like deformations on the trunk, near the grafting site, is often the result of incompatibility of rootstock and scion. In stone fruit trees, this may assume the form of cracks at the graft union, in the bark as well as the wood. This leads to the die-back of branches, with resin exuding from the main branches. Trees may die at various ages. Leaf margins sometimes roll upward.

References

Anonymous (1968) Plant disease development and control. Natl Acad Sci, Washington DC, Publ No 1596

Ausher R, Shoham Ch (1977) Identification of pests and diseases in field and vegetable crops. Isr Minist Agric, Ext Serv

Grogan AR (1981) The science and art of plant disease diagnosis. Annu Rev Plant Pathol 19:333–351

Hicks DR (1976) Diagnosing vegetation injury caused by air pollution. US Environ Protect Agency, Washington DC

Johnston A, Booth C (eds) (1983) Plant pathologist's pocketbook. Commonwealth Mycol Inst, Kew, Surrey, UK

Levitt J (1972) Responses of plants to environmental stress. Academic Press, London New York

Mansfield TA (1983) Effect of air pollution on crops. In: Johnston A, Booth C (eds) Plant pathologist's pocketbook. Commonwealth Mycol Inst, Kew, Surrey, UK

Sprague HB (ed) (1964) Hunger signs in crops. McKay, New York, 461 pp

2.2.2 First Steps in the Detection and Diagnosis of Pest and Pathogen Trouble

Z. ARENSTEIN, J. GOLAN, and J. PALTI

The symptoms of pests and diseases attacking crops are much too varied to be described usefully in short comparative schemes. Keys for the diagnosis of such symptoms on major crops are available, e.g., in the excellent compendia on diseases currently issued by the American Phytopathological Society; the compendia so far published include diseases of wheat, barley, maize, cotton, groundnut, potato, soybean and lucerne.

There are, however, some criteria uniformly valid for recognition of pests and pathogens on numerous crops. These refer in particular to the pattern of symptom distribution on the affected plant or field and to the botanical range of the plants affected. As has been noted above, it is typical of the great majority of pests and diseases that they initially attack isolated plants or small groups of plants, and form foci of infection before they spread over whole fields or orchards (see Sect. 2.2.1.2). The exception are large-scale infestations by invaders from distant sources, especially by locusts and by some moths, or damage caused by migrating birds.

We shall here discuss some symptoms caused by pests or pathogens, first in general terms applicable to many field and orchard crops, and then more specifically in relation to each one of these two groups of crops. This involves a certain amount of repetition but is meant to facilitate the consideration of these symptoms as a first approach to diagnosis. Examples have largely been drawn from symptoms observed in the field and orchard crops grown in subtropical areas, with which the authors are most familiar.

Field diagnosis of diseases caused by pathogens is limited to those which exhibit symptoms clearly recognizable by the grower or adviser. In all other cases, the services of plant clinics or expert phytopathologists are needed.

2.2.2.1 Symptoms Caused by Pests and Pathogens on the Host as a Whole or on Major Parts of the Host

Two groups or categories of parasites affect the host plant as a whole, or concurrently affect major parts of the host: agents of virus, viroid, or mycoplasma diseases, and parasites injuring the root system, root collar, or main stem near ground level (see Fig. 1).

Distinguishing between these two groups is usually not difficult. *Virus, viroid, and mycoplasma* diseases produce varied symptoms on all or most aboveground parts of the affected plant, but the host is killed outright only in certain exceptional, though important, cases (see below). Although these diseases are mostly systemic and the sap of all parts of the plant carries the pathogen, their symptoms appear earlier, and are more conspicuous and severe on the new growth, while older, mature foliage may appear healthy.

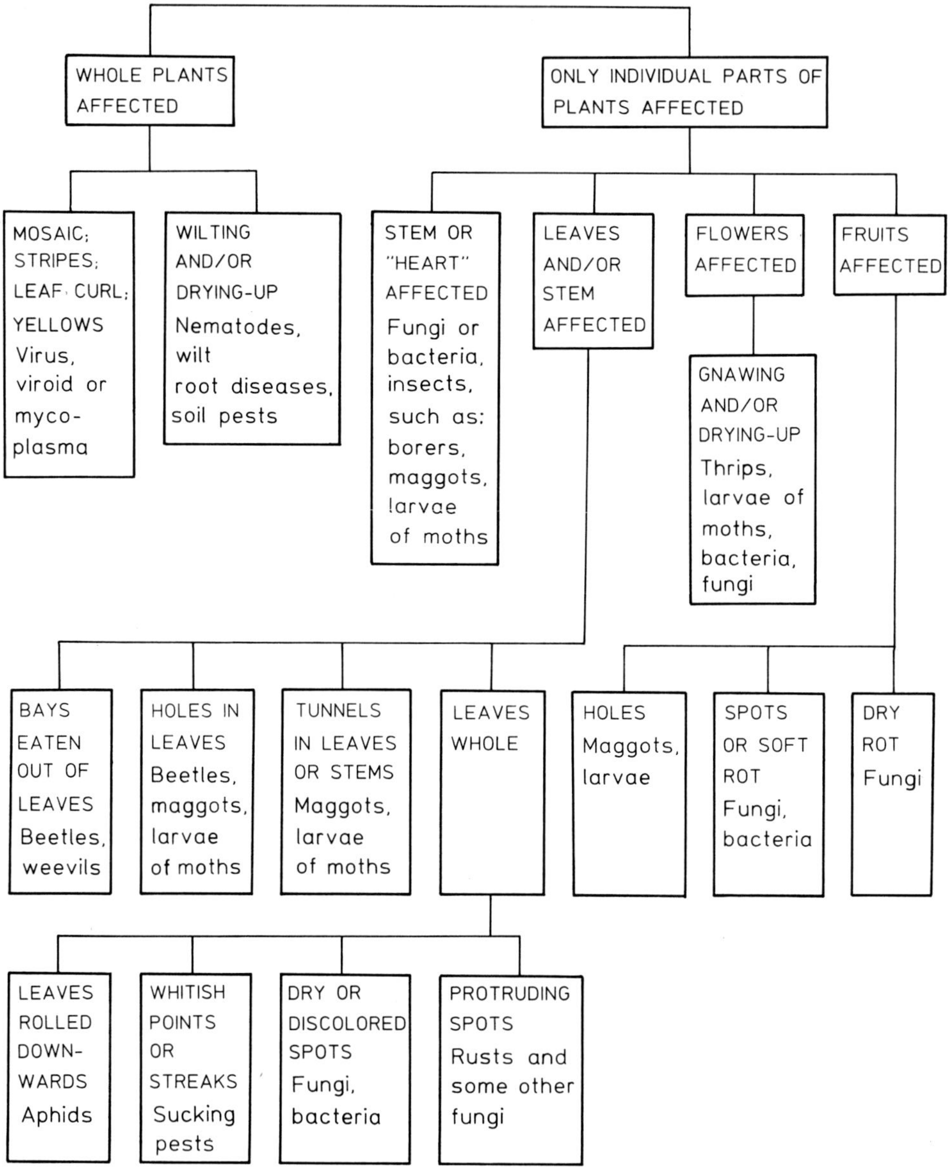

Fig. 1. Key to some symptoms and their most common parasitic causes, where only *single* plants or *small foci* within field crops are initially affected. (After Ausher and Shoham 1977)

Root afflictions, whatever their cause, almost inevitably lead to progressive wilting and drying up of plants, but the rate at which this takes place varies widely. If nematodes are the cause, most affected crops show gradually increasing signs of stunting, before the crop eventually succumbs. Root-knot nematodes are readily diagnosed by the nodules they form on roots, but other nematodes are harder to recognize, although they cause extensive damage, e.g., on citrus and banana.

Most fungi, bacteria, and insects that attack roots of crops kill their herbaceous hosts quickly, and this is sometimes preceded by changes in leaf colour. But in tree crops, the time elapsing between first infection and eventual death may span several years. If stems of wilting or drying plants are cut open and their xylem exhibits brownish or reddish colouring, this generally indicates fungal or bacterial diseases such as those caused by species of *Fusarium* or *Verticillium* or of the *Pseudomonas* and *Xanthomonas* group of bacteria.

2.2.2.2 Common Symptoms of Pests and Pathogens Affecting Single Organs of the Shoot

From the great variety of symptoms pests and pathogens may cause on individual leaves, branches, twigs, flowers, and fruits, we have selected arbitrarily some of those commonly found and characteristic of the effect on the host of certain injurious insects, mites, fungi and bacteria (see Fig. 1).

Spots on Leaves and Fruits. A distinction must be made between dry, brownish spots, light-coloured or whitish spots, spots surrounded or covered by fungal growth, and slightly raised, small spots.

Where *dry* spots are encountered, the first step in diagnosis is to make sure that they have not been caused by hail or chemicals (see Sect. 2.2.1.2). Dry, brownish spots are commonly symptoms of fungal or bacterial pathogens. If such spots are wholly or partly covered by external growth (mycelium and conidiphores or conidia), if they are surrounded by such growth, or if they show concentric rings, they are certain to be due to fungi. The colour of such external growth can serve as a clue to the identity of the pathogen, especially if this is a downy mildew or allied fungus. Most dry spots thus produced are visible on both surfaces of the leaf, but the external growth covering or surrounding them is limited to the lower surface of the leaf. The exception are symptoms of powdery mildew, which appear on both surfaces of the leaf and sometimes on terminal shoots of woody host plants, and these mildews take longer than most other leaf fungi to make affected tissue dry up. Black, superficial spots, often covering much of the leaf or fruit, are due to sooty mould fungi developing most commonly on the excretions of various sucking insects, especially aphids and white flies.

Light-coloured or *whitish* spots or streaks on leaves and fruits are frequently caused by sucking pests, including thrips, leafhoppers, and mites. Various stages of these pests or their moults are generally found on affected organs.

Spots marked by small protuberances are due to fungi. The most common are rusts which form pustules of varying colours, yellow, brown, or black, on leaves and green stems, but rarely on fruits, the exception being legume pods and some deciduous fruits (apricots).

Holes and Tunnels. Holes appear on leaves in many forms and sizes. On tender plant tissue, holes may often be due to rupturing, as the leaf expands, of tissue injured by sucking insects at an early growth stage, as observed on cotton seedlings affected by thrips. But most holes are caused by gnawing insects or slugs. The size of the hole may afford a clue to the size of the causal agent: fleabeetles cause small holes, and so do very young larvae of many moths and beetles. Large holes in leaves are caused by older larvae of these pests, and these also bore into fruits and other organs. Bays eaten out of leaf margins are due to adult beetles and weevils. Leaf tunnels, in which most of the leaf tissue has been eaten up, but the lower and/or upper epidermis persists, are most frequently due to maggots. Such leaf tunnels often contain the blackish excrements of these pests.

Holes not exceeding a few millimeters in diameter are caused by a number of leaf fungi, e.g., *Cercospora beticola* on beets and rusts of stone fruits; such holes are often ringed by dark leaf tissue.

Holes and tunnels in woody parts of the crop will be discussed in the section on diagnosis of pests and diseases in orchard crops (Sect. 2.2.2.4).

Dry and Soft Rots. All these rots are caused by fungi or bacteria, acting either as primary parasites or frequently as secondary parasites infecting the crop after it has first been injured mechanically, by pests, or by any other agency. The most common causes of dry rots are fungi, while bacteria are somewhat less common in this role. On the other hand, soft rots are most commonly due to bacteria, the action of which results in the watery lysis of tissues.

Crinkling or Curling of Leaves. If not due to virus infection, this is mostly caused by pests or pathogens affecting young leaf tissue. Foremost among these are thrips, leafhoppers, mites, and especially aphids, and they are relatively easy to detect. Some fungi attacking young leaves also make them curl, the best-known cases being those of leaf curl on stone fruits (*Taphrina deformans*) and some downy mildews on vegetable crops and tobacco.

2.2.2.3 Pest and Pathogen Symptoms Common on Field and Vegetable Crops

Pre- and Post-emergence Afflictions. Where seedlings fail to emerge, and this is with certainty not due to inferior seed quality, excessively deep sowing, or soil crusts, it must be ascertained whether the seeds have disappeared or have been affected before or after incipient germination. Disappearance of seeds is commonly due to birds or rodents, which leave behind traces that are fairly easy to recognize. Seed rotting is caused by numerous fungi, most of which belong to the damping-off complex, unless extreme environmental conditions prevail. This should in general not be too much of a problem if the seed has been disinfected properly. But hollowing out of seed and browning or gnawing of germinating seed is often due to slugs, wireworms, and various seed-infesting flies, such as *Hylemyia cilicrura*.

Post-emergence injury to seedlings is mostly due to birds and rodents, which pull up the young plants, or to slugs, wireworms, cutworms such as *Agrotis* spp., chafer grubs and other insects that bite seedlings off near soil level. Blackening and rotting of the seedling at or below the collar is a sign of damping-off, caused singly or jointly by species of *Fusarium, Rhizoctonia solani, Macrophomina phaseolina* and other soil-borne fungi.

"Heart Trouble". The wilting or drying up of the center of apical growth is frequently caused by pests with a preference for the tender tissue of such growth. In cereals this may be similar to the injury by stem-boring larvae of Lepidoptera, but the symptom is also often caused by maggots, such as those of *Atherigona* spp. commonly affecting sorghum. In many other crops injury to the "heart" of the plant is caused by the larvae of moths, e.g., by species of *Phthorimaea* on beet and potato, and by *Hellula undalis* on Brassica crops.

The Host as a Whole Is Affected (Fig. 1). On field and vegetable crops, attacks by virus, viroid or mycoplasma diseases are characterized by symptoms of mosaic (pepper, tomato, potato, cucumber and many other crops); stripes, often observed on cereals; leaf curl, common on potato and tomato; yellows, widely distributed on sugarbeet and legume crops; pronounced stunting; and various colour changes and fruit deformations. Less frequently, these diseases cause destructive desiccation of plant organs, leading to the death of the plant.

Pest or disease attack on the roots of field crops can sometimes be diagnosed early by changes in leaf colour, which show up a considerable time before the plants begin to dry up. Thus, potatoes affected by wilt-inducing species of *Fusarium* show reddish tinges in their top leaves, while brownish areas appear on leaves of cotton and solanaceous crops affected by *Verticillium,* and bluish-grey tinges on leaves of Brassicae attacked by the cabbage root fly (*Chortophila brassicae*).

Injury to the whole plant may also be caused by stem-boring larvae which make plants break off near soil level, an injury often observed in cereal crops, especially in maize and rice.

Individual Organs Are Affected (Fig. 1). The range of symptoms on leaf blades has been described above (Sect. 2.2.2.2). In addition, some fungi and bacteria cause affected leaves of dicotyledonous crops to drop prematurely, and this may be the most noticeable symptoms once disease development is advanced. Examples are the powdery mildew (*Leveillula taurica*), which causes extensive leaf drop of pepper (but not of tomato and other hosts), and *Alternaria macrospora* inducing leaf drop on cotton. Small protruding spots have been mentioned above as often due to rusts, and pustules of these fungi are very common on a wide range of field crops. Small, usually black, protruding bodies the size of a pinhead are often observed on dry spots caused by fungi, e.g. the pycnidia formed by species of *Septoria* on cereals and many other hosts.

2.2.2.4 Pest and Pathogen Symptoms Common on Orchard Crops

The Tree as a Whole Is Affected. Virus, viroid and mycoplasma diseases on orchard crops find expression in a great variety of symptoms. On grape-vine, viruses and viroids causing mosaics or leaf deformation ("fan-leaf") are widespread. On deciduous fruit they cause a diversity of symptoms, including mosaics, yellows, "witches broom" and stunting. On subtropical and citrus fruit, these diseases may also find expression in symptoms additional to the above: the sunblotch viroid of avocado causes yellow-red, wide stripes on the fruit, and on citrus trees that suffer from xyloporosis, pitting of the stem is found, if the bark is removed. Similar stem-pitting is also caused on deciduous trees and grapes by certain viruses. An important exception to the general rule that virus, viroid, or mycoplasma diseases are rarely lethal to their tree hosts is constituted by the tristeza disease of citrus, which first kills the peripheral parts of the tree (tips of twigs and roots) and then the whole tree.

As for root diseases, on tree crops the period elapsing between first attack of parasites on the root system and eventual death of trees may be much longer, and often spans several years. In these crops, prompt and correct diagnosis in some cases provides a chance for curing the tree, or at least for prolonging its life considerably. The speed with which root diseases make the tree decline is therefore a factor of great practical importance.

Gradual decline and leaf drop characterize diseases like that caused on avocado by *Phytophthora cinnamomi,* of *Verticillium* fungi on stone fruit and olives, where curative measures can be taken (Vigouroux 1975), and of pests such as *Phylloxera* on grapevine. On the other hand, many root-rotting fungi, including *Sclerotium rolfsii* and *Armillaria mellea* on deciduous and other trees, cause sudden death, and so do various species of root- or stem-boring insect pests. But some such borers found on declining trees are the result rather than the cause of the decline. Thus certain scolytid beetles bore holes into trunks and branches of deciduous fruit trees after these have been weakened by environmental factors.

Individual Organs Are Affected. As far as dry spots on leaves or fruits are concerned, it must be repeated here that the first step in diagnosis is to exclude the possibility that such spots are due to hail or chemical scorching. When this has been done, the range of pests and pathogens that cause spots, as described in Section 2.2.2.2 has to be considered, to distinguish between spots of various colours, with or without covering or surrounding fungal growth, etc. Among the dry spots caused on fruit, those produced by the scab diseases of apples and pears (*Venturia* spp.) are of great importance: these spots are greyish-black in colour, up to 1 cm in diameter, and often develop cracks.

Leaf drop due to root diseases has been mentioned. But this is also a conspicuous symptom of many leaf diseases: olive trees attacked by *Cycloconium oleaginum (Spilocaea oleagina)* or stone-fruit trees attacked by rust (*Tranzschelia* spp.) may be left bare of all except a few leaves at the top of their branches.

As in field crops, small, black, protruding bodies forming on dry leaf spots may be due to fungi such as *Phyllosticta viticola,* the cause of an important disease of grape vine.

Holes and tunnels in branches and limbs are of particular importance on tree crops. Most are due to boring and tunneling beetles, grubs, and larvae, and the points of entry are often marked by excreta and gnawed tissue ("Frass"). Affected branches eventually die off, as can readily be seen in apple branches into which larvae of *Zeuzera pyrina* have bored.

Cankers, i.e., lesions of rotting tissue, sometimes in discrete spots, often girdling the affected part wholly or in part, appear on twigs, branches, limbs and trunks of many tree crops. These lesions may extend deep into the affected tissue. When the rotted tissue is dry, the canker is caused by fungi, e.g., the *Nectria* canker of pome fruit and other trees. But if gum exudes from the canker, the cause may be a fungus, such as species of *Phytophthora* on citrus, or a bacterium, e.g., *Pseudomonas syringae* on stone fruit.

Dry and soft rots, due to fungi and bacteria, are extremely common on fruit crops. Many of the pathogens responsible attack various organs of the tree, e.g., *Diplodia natalensis* and *Phytophthora citrophthora* affect both fruits and woody parts of the citrus tree. In such cases, diagnosis must be comprehensive and consider all the organs potentially susceptible.

There are many types of soft rot. Some of the pathogens involved are mostly restricted to ripening fruit, e.g., the greyish rots caused by *Rhizopus*, or the brown rots due to *Monilia.* Watery rots are often caused by bacteria, which generally penetrate through wounds such as insect punctures or lesions caused by hail, wind, or during picking. Some fungi, including *Rhizopus*, can also use such wounds as points of entry to produce fruit rots. Some other soft rots are characterized by formation of concentric rings on fruit affected, as seen on apple fruit infected by *Gloeosporium* spp.

2.2.2.5 Diagnosis of Combined Stress and Pest or Pathogen Afflictions

The distinctions between non-parasitic afflictions and those caused by pests or pathogens breaks down where unfavourable environmental, nutrient, or other factors cause stress that predisposes the crop to the attack of weak parasites, which would not otherwise be capable of affecting it. The difficult task facing the diagnostician is then to determine what is the primary and what the contributory cause of the affliction, the stress factor or the parasite.

Wallace (1978) has reviewed methods of diagnosing plant diseases of such complex etiology. He cites an example that aptly illustrates the problem: in S. Australia, wheat that failed to develop was found to suffer primarily from lack of available water, and the presence of the take-all fungus (*Gaeumannomyces graminis*), of *Rhizoctonia solani*, and of two species of nematodes on the roots constituted a contributory factor only.

Environmental stress factors often combine with the stress of advancing age to make maturing crops particularly susceptible to attacks by weakly parasitic fungi and bacteria that cause moulds and rots. A case in point is the charcoal rot (*Macrophomina phaseolina*) of maize, sorghum, and many other crops. Identification of the stress factor is therefore of special importance in crops approaching harvest.

On fruit trees, diagnosis of weak pests, such as scolytid bark beetles, is comparatively easy. The difficulty often lies in determining the factor that has weakened the trees to predispose them to the beetles' attack.

The best way to diagnose disorders of complex nature is often to conduct comparative surveys of several or even numerous fields or plantations. The survey should comprise crops of a variety of ages, topographical locations, soil types and cultural practices (density of sowing, nutrition, mode and timing of irrigation, sanitation, pruning, etc.), and record the nature and timing of treatments with pesticides, fertilizers and growth substances.

Reliable diagnosis of crop diseases of complex etiology often requires, in addition to surveys, field experimentation to evaluate the contributory role of some of the factors involved. This, as Wallace (1978) has noted, can help to develop regression models for particular crops, to identify their major constraints in a given area.

2.2.2.6 A Look into the Future: Computer-Based Disease Diagnosis

Research on the potential use of computers for the diagnosis of crop diseases has been initiated in the U.S.A. around 1980, shortly after the publication of similar work on medical diagnosis. Michalski et al. (1983) have investigated such use of computers for the diagnosis of soybean diseases in Illinois. The study covered 17 non-parasitic disorders and infectious diseases. It was based on a series of questions, the answers to which progressively narrow the range of diseases that could be implicated. Thus, to diagnose the powdery and downy mildews of soybeans, a list of questions was drawn up, which includes:

– Time of occurrence:	August or preceding or following months
– Precipitation:	Normal or abnormal
– Temperature:	Normal or abnormal
– Damaged area:	Whole fields or parts of fields
– Leaf spots:	With or without yellow halo
– Mildew growth:	On upper or lower leaf surface
– Premature defoliation:	Present or absent

It is to be hoped that continuation of these studies will facilitate diagnosis of crop diseases and pests in the future, where suitable equipment can be made available.

References

Ausher R, Shoham Ch (1977) Identification of pests and diseases in field and vegetable crops. Isr Minist Agric, Ext Serv

Michalski RS, Davis JH, Bisht VS, Sinclair JB (1983) A computer-based advisory system for diagnosis of soybean diseases in Illinois. Plant Dis 67:459–463

Vigouroux A (1975) *Verticillium dahliae*, agent d'un déperissement de l'olivier en France. Ann Phytopathol 7:37–44

Wallace HR (1978) The diagnosis of plant diseases of complex etiology. Annu Rev Plant Pathol 17:379–402

2.2.3 The Disease Identification Service or Plant Clinic

I. DISHON and J. PALTI

2.2.3.1 Function and Scope of the Disease Identification Service

Diagnosis of crop disorders and diseases in the field concentrates on:

a) Distinction between symptoms of physiological disorders and of pest and pathogen attack.
b) Attempts to disentangle the often complex relationships between host stress and pathogen pest and action (see Sect. 2.2.2.5).
c) Identification of pests and pathogens by their symptoms or, in case of pests, by physical appearance of their various stages.

Such identification, by experienced farmers and advisory staff, can go a long way where routine diagnosis of fairly familiar pests and diseases is concerned, especially when symptoms are observed on aboveground parts of the crop.

The identification service is chiefly concerned with all that is required to supplement field diagnosis. Since insects are more readily diagnosed in the crop than fungi, bacteria, nematodes, and viruses, these pathogens usually claim most of the attention of the plant clinic's staff. The range and accuracy of such identifications depends on the expertise of the staff. In some countries, such as the U.K., this includes research phytopathologists, nematologists, and entomologists who can perform identifications in their own laboratories (see Chap. 3.1); in other countries the clinic acts chiefly as a clearing house for distributing all but the most easily diagnosed samples to universities or other research institutions.

Plant clinics operate in most states of the United States and their work has been reviewed by Evans-Rule (1982). She found that no two clinics are exactly alike, because growers' requirements and funding (mostly by universities) differed so widely.

The work of the plant clinic in many ways parallels that of a hospital's admission ward. It consists primarily of:

1. Winnowing easily diagnosable afflictions from those requiring more intensive study.
2. Indicating, where possible, emergency treatments to prevent aggravation of the affliction and potential reduction of yield.
3. Referring samples with afflictions not easily diagnosed to further examination by appropriate experts.

Among the additional tasks that can be imposed on the plant clinic is that of establishing, by long-term collection of records, a data base to indicate the spatial and seasonal distribution of disorders, pests, and diseases in the country or in a given region. This has served, in Israel, for the empirical determination of "threshold dates" before which certain pests or diseases are not expected to appear (see Sect. 2.4.2.2). However, in many cases the frequency with which a pest or pathogen appears in the samples submitted to the clinic is not necessarily a measure of that parasite's occurrence. As pointed out by Fletcher (personal communication), the most common problems may be recorded infrequently at the clinic, because farmers and field advisers can identify them themselves; enquiries in such cases are a reflection of the unusual, and often perhaps the insoluble, problems encountered in crops.

A service recently developed at the clinic in Israel's Negev district is the quantitative determination of the size of pathogen inoculum in plots farmed intensively. The grower himself takes soil samples according to instructions issued by the clinic's staff, and the presence or absence of pathogens in these samples is determined against nominal payment. This helps growers to plan their crops and, in particular, to decide whether or not soil disinfection treatments are required.

The benefit to be derived from plant clinics depends largely on the training and experience of their staff and on the facilities placed at their disposal; even where the clinic's function is essentially limited to distributing samples for diagnosis to experts, this requires discernment and experience. But the usefulness of clinics also depends greatly on the understanding of those that submit samples for diagnosis, especially the advisory field staff, of how the clinic works, where are its limitations, and of the supreme importance of submitting samples with fullest possible details of previous and current crop history (see below).

2.2.3.2 What Crop Afflictions Most Frequently Require Identification by the Plant Clinic?

How many and what types of crop affliction the adviser refers to the plant clinic depends obviously on his experience and judgement. In a very general way, these types may be classified as follows:

1. Border-line cases of physiological and pest and disease affliction (see Sect. 2.2.2.5). The nature of the parasite has to be diagnosed, whether it only takes advantage of environmental deficiencies in the crop (drought, lack or excess of nutrients, etc.) but is of little importance once such deficiencies have been corrected, or whether it can be the primary cause of loss. The physiological component of the crop's ailment cannot usually be diagnosed in the clinic. An example of what is frequently such a border-line case is the charcoal rot of maize (*Macrophomina phaseolina*). Where this fungus is associated with rotting, the basic cause is often found to be drought and heat (Dhingra and Sinclair 1978).
2. New or rare pests and diseases, the symptoms of which are unlikely to be familiar to the adviser. Where new pathogens are concerned, the work of the plant clinic may have to include, if possible, demonstration by inoculation tests of the extent to which the pathogen can serve as primary cause of infection and loss.

3. Organisms, especially fungi, that have to be coaxed to reveal features by which they can be identified (Spaar et al. 1984). This includes fungi affecting aboveground organs, if they can be induced to form identifiable fruiting bodies (e.g., powdery mildews induced to form cleistothecia), but refers more particularly to fungi and bacteria affecting belowground organs. Such pathogens often, on affected seedlings, roots, and collars, cause symptoms hard to distinguish from one another; their true nature can be determined only by culturing on selective media. The best example is probably the damping-off of seedlings, which may be caused by many fungi, including *Rhizoctonia solani,* and species of *Pythium* and *Fusarium,* acting alone or in complexes, and identification of the fungus responsible (or dominant in the complex) can be important for the further development of the crop and of succeeding crops.
4. Virus diseases and nematodes. Not many clinics are equipped to carry out all stages of the work required to identify viruses, and expert virologists usually have to be called upon to do this. The same applies, though to a lesser degree, to nematodes, especially to the identification of free-living nematodes.

2.2.3.3 How Can the Adviser Derive Full Benefit from the Clinic?

The output of the clinic is only as good as the input by its clients, in many cases the adviser, i.e., the quality of the samples submitted. The answers to be provided by the clinic are, mostly, only qualitative (*what* causes the trouble?) and leaves quantitative assessment and prognosis to the client, especially answers to the questions: *how much* of the field is affected, and *how fast* is the trouble likely to spread?

The most important duty of the adviser or grower submitting a sample to the clinic is to ensure that:

a) The sample is typical of the afflictions to be diagnosed.
b) The sample is packed most carefully to make sure it arrives in the clinic in a state fit to be examined.
c) Fullest details of the crops are given, including:
   - previous crops grown in the two preceding seasons
   - crops bordering the affected field
   - type of soil and topography
   - date of sowing or planting and crop density
   - cultural practices, especially fertilization, mode and frequency of irrigation
   - chemical treatments applied, including pre-plant soil disinfection
   - frequency and pattern of disease in the field
   - unusual or extreme weather prior to pest or disease appearance.

In sum, integration of work of the advisers and the clinic's staff is of the greatest importance. For this purpose, the adviser should at times participate in the clinic's work, and the staff of the latter should from time to time join advisers in their farm visits. The introduction of mobile plant clinics in some U.S. states is a welcome development. Such clinics help in training advisers, and sometimes also growers, in diagnosing problems and in understanding how the clinic works.

References

Dhingra OD, Sinclair JB (1978) Biology and pathology of *Macrophomina phaseolina.* Imp Univers Univ Fed Viscosa, Brazil, 166 pp

Evans-Rule G (1982) Plant disease clinics – past, present and future. Plant Dis 66:80–86

Spaar D, Kleinhempel H, Fritzsche R (1984) Diagnose von Krankheiten und Beschädigungen an Kulturplanzen. In: Wetzel T (ed) Diagnosemethoden. Springer, Berlin Heidelberg New York, 198 pp

2.3 Crop Value, Economic Damage Thresholds, and Treatment Thresholds

J. PALTI and R. AUSHER

The crop protection adviser is constantly faced with the problem when to counsel the farmer to act to mitigate or prevent damage from abiotic or biotic agents, either by strengthening the plant (water, fertilizers) or by checking pests or pathogens by chemical or cultural means. This is an economic decision taken against the background of biological and ecological considerations: will it, or will it not, benefit the farmer and increase his income if he goes into action to minimize risk or save some of his yield, and can he do so without deleterious effects on pest predators or parasites, or on bees, fishes, wildlife etc., and without undue health hazard (Diercks 1984)?

Cook (1985) has recently stressed that negative effects of pest or pathogen on yield should not be referred to as "loss", because, as he puts it "farmers cannot lose what has never been produced in the first place". Yield not realized by the crop owing to interference of disorders, pests, or pathogens should be termed "yield reductions", and this term will be used here.

In this chapter we shall first consider the factors that determine reduction of yield and income, and the extent to which such reduction can be mitigated or prevented. We shall then proceed to discuss the yield reduction preventable by farmers' operations and distinguish between two types of threshold:

a) the economic damage threshold, which is the point at which any factor, whether biotic or abiotic, begins to reduce the value (quantity and quality) of the yield; and
b) the treatment (control) threshold, which is the point at which the value preventable by farmers' operations exceeds the cost of control operations, i.e., the extra yield obtainable by such operations promises an income higher than the control expenses. Criteria for establishing treatment thresholds have been discussed by Carlson and Main (1976).

2.3.1 Factors Affecting the Level of Yield Reduction and the Economic Damage Threshold

Crop growth is subject to many and varied constraints, many of which have no measurable effect on the size of yield. Definition of the point at which such constraints cause actual loss in yield, i.e., the economic damage threshold, is complex and requires experience. It involves prognosis of certain weather factors and of crop and pest development.

Numerous factors contribute to the fixation of damage thresholds, including seasonal considerations and market aspects. The following crop-related factors may be considered of importance.

a) Combined effects of several abiotic factors, such as heat and drought, of several biotic factors, such as pests injuring and bacteria rotting fruit, and interaction of abiotic and biotic factors, such as combined stress and parasite damage. An example of the latter, as mentioned in Chapters 2.3 and 2.4, is the enhanced effect of charcoal rot (*Macrophomina phaseolina*) on maize yield under hot and dry conditions.
b) The proportion of marketable yield *directly* affected. Thus moderate infestation of olives by the olive fly (*Dacus oleae*), which affects the fruit directly, may be more damaging than more severe incidence of the olive peacock leaf spot (*Cycloconium oleaginum*), which infects leaves and has only an indirect effect on yield.
c) The chances the crop has to recover from injury, or to compensate for such injury. These chances depend, inter alia, on the amounts of water and nutrients available to the crop, on impending changes in weather (such as transition from rain-free to rainy season or vice versa), and on the growth stage of the crop. Injuries suffered at different growth stages may differ very widely in their impact on yield. In particular, injury to tender stages (seedlings, blossoms) or close to maturation may leave the crop few chances of recovery. In this connection, great value attaches to the detailed definition of growth phases (stades repères) worked out in various countries, for instance those proposed by French horticulturists for many fruit and some field crops (Anonymous 1952) as discussed in Section 2.4.2.4. The crop's capacity to compensate for injured organs or for missing plants also depends on its growth habit: the cotton plant, which tends to spread and fill available space, and winter cereals that tiller, make up for missing plants more readily than sugar beet or maize.
d) Interference with harvesting operations (lodging, weed growth).
e) The purpose for which the crop is to be marketed, e.g., for canning or for fresh sale. Stern (1973) cites the example of corn grown in Canada and affected by the European Corn Borer (*Ostrinia nubilalis*): canning factories established a tolerance of 10% ears affected by the pest, while not more than 5% of such ears were acceptable for fresh market corn.
f) Potential damage to neighbouring and/or succeeding crops.

2.3.2 Damage or Yield Reduction Amenable to Mitigation or Prevention

Damage that can neither be mitigated nor avoided has to be considered "force majeur" and the only decision the farmer has to take in such cases is whether or not the crop is worth keeping in spite of the damage caused. This means in effect weighing the expenses listed in the previous section against the reduced income to be expected from the damaged crop. This sort of catastrophe is comparatively rare, e.g., frost destroying bananas or citrus, or locusts devastating crops at a stage of growth at which recuperation of the crop is unlikely.

But where some of the damage already caused can be mitigated or future damage can be prevented, farmers have to take the complex decision whether treatments designed for this end will be profitable (see next section). This involves as accurate as possible an appraisal of what, for the sake of brevity, we shall here refer to as "preventable" damage.

Amenable to mitigation is much of "one-time" weather damage (chilling, hot winds, hail, and sometimes frost), from which plants can be helped to recover or for which they can compensate. Mitigation of damage is also possible for certain parasites, especially those affecting the crop gradually, with no direct effect on the yield; thus loss caused by root-knot nematodes can often be lessened by application of nitrogen.

Assessment of the chances of preventing damage involves prognosis of future pest and disease development and of the means available to reduce both the rate and ultimate size of such development. The aspects to be considered here include:

a) Dynamics of pest and disease multiplication, with due regard to:
- Proximity and size of sources of infestation, such as clumps of potatoes affected by *Phytophthora infestans*, neighbouring crops, groundkeepers or volunteer plants, etc.
- Infestation potential or vitality of pest: if the pest population is nearing the end of its harmful phase (larvae near pupation), or if it is parasitized or subject to predators or enzootic diseases, avoidable damage may be less than that from vigorously multiplying pest populations.

b) Susceptibility and growth vigour of the crop: a vigorous crop can make up for damage caused, or may outgrow pests or pathogens, and this increases the amount of damage that is preventable. However, some crops are most susceptible when vigorous (e.g., cereals affected by rusts). Where one particular stage of growth is threatened (seedling or blossoming stage), damage may largely be preventable by accurately timed measures, but once damage has been done at this stage, there may be little room for prevention at subsequent crop stages.

c) Availability and efficiency of monitoring services: the extent to which damage can be prevented changes as crops and pests pass through the various stages of their development. Assessment of preventable damage with any degree of accuracy requires constant monitoring.

d) Pesticide-related aspects:
- Availability of suitable pesticides, labour and machinery for application: if the pesticides available are not those most desirable, if labour is not adequately trained, and if the choice of machinery is strictly limited, the amount of preventable damage diminishes. Thus, if pest or pathogen damage could be prevented by prompt treatment, but the soil is wet, damage could be averted by aircraft treatment, or perhaps by knapsack-spraying, but not where only tractor-drawn or self-propelled heavy ground equipment is available.

- Efficacy of pesticides against various phases of pests: many pesticides are limited in their effect to certain phases of pest development, e.g., they may be active against larvae but not against eggs, against young but not against fully grown larvae, against fungal spores but not against sclerotia. The range of phases against which the pesticide can act will often determine the amount of preventable damage.
- Effect of pesticides on beneficial organisms, such as parasitic or predacious insects or mites, or antagonistic or competitive soil fungi. Where the preventive action of the pesticide is additional to that of beneficial organisms, preventable damage will evidently be greater than where the pesticide limits or abolishes the action of the latter (see also the following sections).

2.3.3 Prognosis of Preventable Yield Reduction

The prognosis of the amount of yield reduction to be expected at harvest consists of fore casts of disease and pest development, which will be treated in Chapter 2.4.2 and of the loss these parasites are likely to cause if left unchecked. This refers solely to that part of the loss which is preventable by control operations.

We have to deal here with the expected effect on revenue not of one agent apt to diminish yield, but of the sum total of such agents, including weeds, rodents, etc. which will not be markedly affected by fungicidal or insecticidal treatments. For example, an insect pest against which effective and cheap insecticides are available may occur in a field liable to attack by rodents that cannot be controlled by the means at hand. Yield reduction that is preventable may then be so small that control operations against the insect pest are not economic.

Prediction of the impact of disease caused by slow-developing pathogens is particularly complex in perennial orchard crops, as has been pointed out by Carter (1978). The number of potential infection courts (e.g., pruning cuts) in such crops increases each year, as the tree grows; the greater the vigour of the tree, the more it has to be pruned, and the number of infection courts grows. But there is a lag phase, often of several years, for disease development. During this phase, host compensation confounds the amount of destruction caused by disease, especially if the grower ably assists the tree in its recovery.

2.3.4 Crop Factors Affecting Determination of Treatment Thresholds

Treatment thresholds for a given crop with a known potential for recovery when attacked can sometimes be determined on a regional scale, if cropping conditions are reasonably uniform and if a single pest or disease stands out as the fulcrum for control operations in the crop. However, more commonly, field-to-field variation will necessitate separate prognoses for each field and orchard, even when these are close neighbours. These variations are primarily due to cultural factors which are more fully discussed elsewhere (Palti 1981), including date of sowing or planting, density of stand, crop vigour, proximity to sources of infection, topography, and irrigation technique and timing.

In addition to the factors mentioned in the two previous sections as affecting the level of potential yield reduction and the extent to which this might be prevented, the following crop factors must be considered before treatment thresholds can be fixed:

a) Composition of the pest, pathogen and weed populations in each field at any one growth stage. This may depend, inter alia, on the health of the propagating material, on the previous history of the field and on several of the above-mentioned cultural factors, especially irrigation.
b) Size and shape of the field. Long and narrow fields are more widely exposed to the influx of harmful as well as beneficial agents from neighbouring fields or uncultivated areas. Where pathogens or pests overseason in such areas, e.g., rusts or *Marsulia* beetles on wild cereals, damage may be expected to be more extensive in narrow than in square fields of moderate or large size. On the other hand, where predators of crop pests breed in uncultivated areas, their influx may lessen the expectation of damage more markedly in the narrower fields.
c) Shelter belts (hedges, rows of trees). Hedges that may shelter insect pests, rodents or birds, are obviously more important to small than to large adjoining fields or plantations. They not only serve as sources of such pests, but also provide shade that may advance many diseases, and act as windbreaks near which aphids and other small flying insects come to earth (Quiot et al. 1979).
d) The period of time likely to elapse between taking the decision to perform control operations and the actual performance of that operation: the longer that period, the greater the amount of possibly unpreventable yield reduction for which allowance must be made.

2.3.5 Economic Factors in the Determination of Treatment Thresholds

The economic factors to be considered before treatment thresholds are fixed, are numerous and varied.

Growing Expenses Already Incurred. The size of investment in the crop, e.g., in seeds, sowing and irrigation, before it has been attacked by pest or disease, may loom large in the grower's mind. To protect this investment he may consider lower treatment thresholds, i.e., to spray even when infestations are relatively low. The investment will be lower, and can more readily be written off, in young crops than in crops nearing maturity.

Sampling Expenses. Field scouting and sampling to appraise the necessity for treatment, involves a fair amount of work. Enlightened growers will consider this an expense similar to crop insurance.

Harvesting and Marketing Expenses. The health status of a crop may affect expenses of harvesting, especially when pests cause crops to lodge or shed fruit, and also has a bearing on the expenses incurred in grading and the potential loss in storage, transit, and sale. Where diseases or pests appearing in field or orchard necessitate post-harvest control treatments, these must be considered in fixing treatment thresholds.

Direct and Indirect Cost of Treatment. This includes costs of water and fertilizer applied to mitigate losses, and pesticides as well as labour employed in their application. The cost of application depends to some extent on the size and topography of the plot and on crop density and soil conditions. Costs may be reduced where application of insecticides and fungicides can be combined with that of herbicides, fertilizers, or growth substances.

Market Price of Produce. Delayed marketing of produce because of the presence of residues applied shortly before harvest could lower the price fetched on the market. On the other hand, better keeping quality due to pre-harvest treatments may make it possible to store the produce until the market price is favourable.

2.3.6 Constraints Affecting Determination of Treatment Thresholds

These constraints are varied in nature and comprise:

a) Potential phytotoxic effects of pesticides, especially when applied in mixtures.
b) Potential damage to beneficial organisms that restrict parasite damage (see also previous section). The threshold for the first treatment to be applied to a crop, when parasites and predators are still active, may be considerably higher than that for a later treatment. Thus, in a case cited by Stern (1973), and referring to control of *Heliothis zea* on cotton in California, the threshold for the first treatment was 15 first and second instar larvae per 100 plants, but for the second treatment this was reduced to only 8 such larvae. Gonzalez et al. (1982), also in California, have developed this concept in what they call the "food-web" approach to the definition of economic thresholds for cotton. They stress the role played by primary as well as secondary pests as a source of food for arthropod predators: where such food is plentiful, the potential for predator development increases, and subsequent treatment thresholds must make allowance for the beneficial action of these predators. Assessment of the sizes of predator and parasite popula-

tions is also an important element in the Integrated Pest Management Programs operated in Texas, U.S.A., as described in Chapter 3.6. The effect of antagonists or competitors of parasitic flora is also to be considered in relation to soil treatments. The partial elimination of such microorganisms facilitates re-infestation, may limit the period over which the treatment brings relief, and thus indirectly raise its cost.

c) Potential enhancement of the development, in the grower's field or in those of his neighbours, of non-target parasites, due to the action of certain pesticides, e.g., increased development of red spider mites after application of some pyrethroid insecticides.
d) Health hazards to operators or to human or animal populations within the range of pesticide drift.
e) Environmental hazards, such as effects on wildlife, fishes and bees, and on neighbouring crops which may suffer phytotoxic damage or reduction of the populations of beneficial organisms.

Points c, d and e may cause the growers (or the operator, who will then devolve the cost on the grower) to incur the expenses of third-party risk insurance, and relate to the regional aspects of crop protection (Chap. 2.4.7).

2.3.7 Use of Computers to Define Treatment Thresholds

Increasing use of computers in the agricultural economy makes it possible to base definition of thresholds on external data such as weather and market prices, and on field data including crop development, estimates, or counts of the sizes of inoculum or of populations of diverse pests and their predators and hyperparasites. Such thresholds can currently be adjusted as the crop develops and conditions change. Numerous such thresholds have been worked out, especially in the United States, and some of those established in Texas are referred to in Chapter 3.6.4. The practical use to which such thresholds have been put in cotton growing in Israel is described in Section 2.4.6.2.

Basing on thresholds worked out by computer, forecasting systems have been developed in various countries. Some of these will be described in Chapter 2.4.2.

2.3.8 Procedures for Defining Treatment Thresholds

The following steps are suggested for the calculation of treatment thresholds to guide control operations:

1. Estimate
a) The general level of yield to be expected from a healthy field.
b) The percentage by which yield is likely to be reduced unless control operations are performed.

c) The percentage of yield reduction considered preventable by control operations, taking into account the growth stage of the crops, the dynamics of development of the harmful and beneficial organisms present or expected, the efficacy of control operations and the frequency at which they have to be repeated, and, last not least, the weather to be expected until the crop matures.
2. Determine (where prices are fixed) or estimate the price at which the yield will be sold.
3. Multiply the amount of preventable yield reduction by the price to be expected, to arrive at a sum expressing the *total gross benefit* to be derived by the control operation.
4. Calculate the sum of all direct and indirect costs involved in the control operation.
5. The treatment threshold has been reached when the total gross benefit to be expected exceeds the sum total of control expenses by a margin considered reasonable by the grower.

The calculation of treatment thresholds thus involves at least four variables, as shown schematically in Table 1.

These calculations take into acount only factors with an immediate bearing on the profitability of treatments, and not the indirect costs mentioned in previous sections. In addition, there are at least three further points the adviser or grower should consider:

a) Where growers find it necessary to carry out control operations not so much to increase profits, but to guard themselves against sudden losses, the insurance value of the operations may make some of the above calculations inapplicable.
b) In the crop loss calculations no allowance has been made for expense *not* incurred for "untaken harvest", e.g., the potential harvest and transport costs of yield that would have been produced if disease had not destroyed it.

Table 1. A sliding scale of treatment thresholds as dependent on four variables: size of expected yield, % preventable yield reduction, price of produce, and level of control expenses

Size of expected yield ($t\ ha^{-1}$)	% Preventable yield reduction	Price of produce ($\$\ t^{-1}$)	Maximum gross benefit from control ($\$\ ha^{-1}$)	Control expenses ($\$\ ha^{-1}$)	Treatment threshold (% yield increase that justifies control)
10	15	6	9	6	10
10	15	6	9	12	None[a]
10	15	12	18	12	10
10	25	6	15	24	None[a]
20	15	6	18	6	5
20	15	12	36	6	2.5

[a] Treatment cannot pay because control expenses exceed the maximum gross benefit obtainable through the control operation

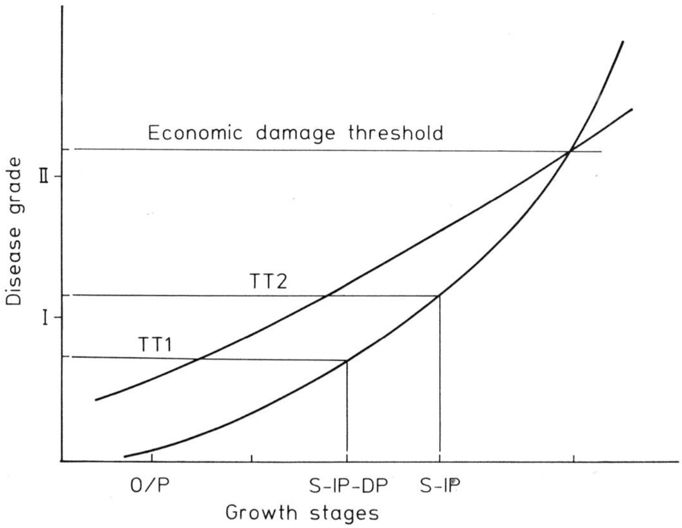

Fig. 1. Determination of treatment threshold according to grade of disease and stage of growth. *S* growth stage; *IP* incubation period; *DP* decision period, i.e., period elapsing between the time the decision to treat is taken and the actual application; *TT1* treatment threshold at growth stage S, minus incubation period IP and decision period DP; *TT2* treatment threshold at growth stage S, minus incubation period IP. (After Kranz and Hau 1981)

c) Many considerations that are not strictly financial may influence the grower's view of a reasonable threshold, e.g., availability of alternate use of labour and storage facilities (see Chap. 2.3.9).

Two approaches to the definition of treatment thresholds are described in the following.

Kranz and Hau (1981) have described procedures for this purpose, and Fig. 1 has been adapted from their work. In this figure, the treatment threshold, at growth stage S, is assumed to lie above grade II of disease incidence. The grower's aim is to prevent disease from reaching or exceeding this grade. The time at which he then has to initiate control operations depends on the disease's incubation period (IP) and the time span he needs to translate his decision to act into actual application (DP). If this span is likely to be longish, the decision to spray or otherwise control the disease must be taken at a correspondingly lower grade of disease (TT1) than when treatments can be applied promptly (TT2).

A practical example how to calculate a treatment threshold has been given by Boethel and Eikenbary (1979) for control of the pecan weevil (*Curculio caryae*) in the United States. The calculation proceeds by the following steps:

1. The rate of oviposition is determined; this has been found to amount to 7.8–9.6, or on average 8.7 nuts per female.
2. The loss in weight caused by a given number of weevils, if left unchecked, is determined; 20 weevils would on average oviposit on 174 nuts. In a variety that averaged 50 nuts per pound, 174 nuts weigh 3.5 lbs.
3. The expected yield per tree has to be estimated, whether it amounts to 100, 200, or 300 lbs annually. If the expected yield amounts to 200 lbs, a reduction of 3.5 lbs caused by 20 weevils is less than 2% of the yield.
4. The monetary value of the yield is calculated. In the example given an expected reduction below 2% was not considered to warrant control.
5. The above reduction in yield was forecast as the damage caused by 20 females. Since the pecan weevil has a sex ratio of 1:1, the grower would have to count at least 40 weevils per tree before the treatment threshold is reached.

It may be noted that in the Soviet Union a simplified across-the-board treatment threshold corresponding to yield reductions of 3–5% in field crops has been postulated (Benedek et al. 1981). The 5% threshold applies to crops with a potentially low-value yield, the 3% threshold to yields of higher value. Since yield in the Soviet Union is frequently sold at a fixed price, calculation of treatment thresholds there is obviously simpler than in countries in which changeable market factors determine the yield's ultimate value.

2.3.9 Treatment Thresholds for Peasant Farmers

Increasing attention has recently been paid to the development of treatment thresholds for peasant farmers by techniques that take into account limitations in literacy, counting, and calculation.

The dominant consideration, where peasant farmers have to decide whether to control parasites or not, is often not the potential loss in revenue, but the extent to which the yield obtained will feed his family until the next crop ripens. Where crops are grown in polyculture, it is unlikely that valid treatment thresholds can be determined. Polyculture tends to limit damage by most pests and diseases on any one crop, and pesticide treatments beneficial to that crop may have direct or indirect adverse effects on other crops present, e.g., by reducing their predator fauna.

On crops grown in more or less pure stands, thresholds can be established more readily, but only if considerations peculiar to the farmer's thinking are taken into account. An instructive example quoted by Farrington (1977) is that of cotton pest control in Malawi. Treatment thresholds based on field experiments were suggested to the farmers. However, farmers eventually adopted thresholds at considerably higher levels because (a) the suggested thresholds involved extra labour for spraying and for harvesting and grading the increased yield; (b) the yield potential, and therefore the amount of preventable damage, was lower than assumed, due to inferior management on most farms; and (c) the risk of poor harvests (which would not repay the cost of treatment) in years of low rainfall was an important consideration.

2.3.10 Where Treatment Thresholds Are Not Considered – Prophylactic Application of Pesticides

Prophylactic or preventive treatments are applied before disease symptoms are apparent, before disease has been forecast to appear, and before pests have been detected by monitoring or traps. Such treatments obviously risk being premature, or even quite unnecessary, and quite apart from needless expenditure, are then liable to harm populations of beneficial organisms, promote resistance of insects, mites or fungi to pesticides, cause outbreaks of non-target pests, and possibly cause phytocidal crop injury.

However, prophylactic treatments are widely practiced, and it is important to see their place in chemical pest and disease control in proper perspective. Thus, such treatments may be essential, justifiable or at least defensible, under one or more of the following circumstances:

a) Where farmers cannot be relied upon to detect initial appearance of pests or diseases in time to prevent reductions in yield, and no monitoring services are available to help them.
b) Where forecasting services are not available, and
   - diseases develop so rapidly that treatments first applied after symptoms have been noticed give inadequate protection, e.g., late blight of potatoes;
   - early disease symptoms are hard to detect even by competent farmers, e.g. downy mildew of onion, *Peronospora destructor;*
   - the marketable produce is directly affected, e.g., by apple scab (*Venturia inaequalis*).
c) For the control of virus vectors.
d) Where the value of the crop is so high that even minimal reduction in yield is unacceptable.
e) In seed-crops or propagating nurseries that have to be kept absolutely free from pests and pathogens specified by law.
f) Where prompt and effective application of pesticides is not assured once diseases or pests have appeared.

Certain diseases, against which prophylactic treatments were commonly practiced when only protectant fungicides were available, can now be controlled by application of systemic fungicides after first symptoms have appeared, and thus become subject to considerations of appropriate treatment thresholds. This includes various powdery mildews, *Botrytis,* and some rusts. However, since frequent use of most systemic fungicides involves the risk of the development of pathogen resistance, they should be used only sparingly and selectively.

Prophylactic treatments are most widely practiced and almost always economically justified where they are applied to propagating material in form of seed dressings, or treatments applied to bulbs, corms, transplants, nursery stock, etc., before these are sown or planted. Such treatments are cheap and involve little outlay for labour. The situation in field and orchard is less clear-cut: here the sometimes considerable cost of one or more prophylactic treatment, and the above-mentioned potential negative biological effects have to be weighed against the "insurance" effect of such treatments and the farmer's wish to be spared as many worries as possible. This is precisely the reason why contract spraying operators engaged by growers for the purpose of keeping crops free from pests and diseases often insist on prophylactic applications, on dates contractually fixed beforehand; they do not wish to risk the charge of having treated the crop too late.

References

Anonymous (1952) Avertissements agricoles. Bull Tech Stat Avertissements Agric, 6 pp

Benedek P, Ebert W, Tanski VI, Dirlbeck J (1981) Aufgaben der Prognose und Methoden der Entscheidungsfindung auf Betriebsebene. In: Überwachung und Prognose – Grundlagen eines gezielten Pflanzenschutzes. Akad Landwirtschaftswiss DDR, pp 220–231

Boethel DJ, Eikenbary RD (1979) Status of pest management programs for the pecan weevil. In: Boethel DJ, Eikenbary RD (eds) Pest management programs for deciduous fruit trees and nuts. Plenum Press, New York London, pp 81–119

Carlson GA, Main CE (1976) Economics of disease-loss management. Annu Rev Phytopathol 14:381–403

Carter MV (1978) Problems of assessment in perennial orchard crops. In: Close RC et al. (eds) Epidemiology and crop loss assessment. Proc Workshop Aust Plant Protect Soc, pp 23/1–23/2

Cook RJ (1985) Use of the term "crop loss". Plant Dis 69:95

Diercks R (1984) Einsatz von Pflanzenbehandlungsmitteln und die dabei auftretenden Umweltsprobleme. Kohlhammer, Stuttgart Mainz

Farrington J (1977) Economic thresholds of insect pest infestation in peasant agriculture: a question of applicability. PANS 23:143–148

Gonzalez D, Patterson BR, Leigh TF, Wilson LT (1982) Mites: a primary food source for two predators in San Joaquin cotton. Calif Agric 36:18–20

Kranz J, Hau B (1981) Wie gewinnt man wirtschaftliche Schadensschwellen? DLG-Mitteilungen 12/81, pp 667–669

Palti J (1981) Cultural practices and infectious crop diseases. Springer, Berlin Heidelberg New York

Quiot JB, Verbrugghe M, Labonne G, Leclant F, Marrou J (1979) Ecologie et épidemiologie du virus du concombre dans le Sud-Est de la France. IV. Influence des brise-vent sur la répartition des contaminations virales dans une culture protégée. Ann Phytopathol 11:307–324

Stern V (1973) Economic thresholds. Annu Rev Entomol 18:259–280

2.4 Pest and Disease Monitoring and Management

2.4.1 Sampling for Crop Pest and Disease Monitoring and Control

2.4.1.1 Sampling of Crops for the Determination of Pest Populations and Treatment Thresholds

A. GENIZI

Taking representative samples is an essential step in monitoring crop pests and diseases. Field sampling methods should be simple, rapid, and adequate in their reliability. More refined techniques may be required wherever higher degrees of precision are needed. The points to be covered in this brief review are the size of sampling plots in fields and orchards, the distribution pattern of sampling units in these plots, the number of units to be sampled, the size of each individual unit, and the evaluation of samples for correct assessment of parasite distribution and frequency. This refers only to direct sampling and not to indirect sampling by trapping.

The principal aim of sampling for pest and disease monitoring, as distinct from sampling for purposes such as yield determination, is not to arrive at a mean value for the crop as a whole, but to determine representative values for each of the reference plots into which the field or orchard has been divided. Each reference plot stands on its own, and decisions taken for or against application of treatments refer to this, but not necessarily to a neighbouring, plot.

Pest and Pathogen Behaviour as Related to Sampling

Sampling methods may not be chosen without prior consideration of the distribution and behaviour of the pest or pathogen at the time of sampling. Two aspects will be mentioned here:

Mobility of the Pest. Sampling, whether it involves detaching plant parts for later examination or inspection of the plants in situ, may cause highly mobile pests to scatter rapidly, e.g., white flies and midges. They can then only be sampled by sweeping techniques (Frankel, pers. comm.). Mobile pests and airborne pathogens invade fields from near-by sources, and may occur in larger numbers on the edges than in the interior parts of the crop. Those parts of the crop that may be endangered first may therefore have to be sampled earlier.

Rate of Spread. Certain pathogens and pests may initially cause severe infections on individual plants or in small loci, long before even minimal numbers of pests or symptoms can be found on neighbouring plants. Detection of such early infections is of the greatest importance and requires taking numerous samples in order to cover the field with an adequately dense "net" of sampling points. On the other hand, where disease spread is rapid, as in the case of potato late blight (*Phytophthora infestans*), the frequency of sampling rather than the number of samples taken must be increased.

Spatial and Statistical Distribution Patterns

Before we approach the subject of the determination of the number of samples and their size, a few remarks on distribution patterns have to be made.

Spatial patterns of distribution can deviate from uniformity in either of two directions, as has been pointed out by Ruesink (1980): if the presence of an individual or infection locus at one point increases the probability of another individual or locus being nearby, then the spatial pattern will be clumped; this, of course, is of frequent occurrence, especially in the distribution of pests of limited mobility (aphids, mites) and of infectious leaf or fruit diseases. Less commonly, presence of a pest may decrease the probability of another being nearby, thus decreasing the frequency of random aggregation of pests.

The statistical frequency distribution depends on both the spatial patterns of infestation and on the size of the sampling unit, as will be discussed later. Where the parasite or symptom spreads at random in the field, and its spread is independent of other parasites or symptoms, the number of individuals to be found at each sampling point will follow what is termed the Poisson statistical distribution. But if a population exhibits an aggregated spatial pattern, and if samples are taken from randomly distributed points in space, the number of individuals per sample will follow one of the so-called contagious statistical distributions, of which the one most commonly found to fit the data in crop protection work is the "negative binomial". Points to be considered in choosing the appropriate statistical distribution in the assessment of parasite populations have been discussed by Ruesink (1980).

Size of Reference Plots

Where large fields or orchards are to be sampled, the required accuracy can only be achieved if they are divided into several parts called reference plots, which will be sampled separately. The size of these plots is an important consideration, which bases principally on the following two aspects:

a) The uniformity of growing conditions, with special reference to factors that may markedly affect pest populations. The more uniform the conditions and the more even the expected spatial distribution of the pest or disease, the larger the plot can be, e.g., from 5 to 25 hectares in field crops, less in orchards. However, even where plant stands appear reasonably uniform, there may be a fairly large variety, and this can be especially meaningful in crop protection work. To obviate errors due to such variability, two measures are advocated:

- Subdivision of the reference plot into subplots (squares) which are themselves more homogeneous than the reference plot as a whole.
- Determination of the cause of variance: this may be a matter of topography (top, middle, or bottom of a slope) or of proximity to sources of infection (other crops, hedges, irrigation points, or other sources). Once the cause responsible has been established, an attempt must be made to define the size of the area over which its influence is felt, e.g., does the influence of the hedge extend only to the rows nearest to it, or perhaps for as much as 20–30 m? This may be exemplified by an observation made in a tomato field near the Sea of Genezareth in Israel. The field was bordered on its southern side by a hedge of cypress trees of varying height, which threw their shade on the field and thus affected the persistence of dew. When *Phytophthora infestans* attacked these tomatoes, the image of the hedge was "photographed" in the field: opposite the taller trees the disease affected many more rows than opposite the lower trees.

b) Various practical considerations also play a role in deciding on the size of the reference plot. The most important of these is the scale on which control operations are to be carried out: if this is a small-scale operation, e.g., plots to be treated by hand-borne equipment, smaller reference plots can be chosen than for fields to be treated by tractor-drawn equipment that may cover many rows in one passage; large reference plots will be chosen if pesticides are to be applied by aircraft, which cannot be selective in spraying some plots, but not others, in the same field. Choosing large reference plots involves ignoring a certain amount of heterogeneity in the pest population.

In dividing the crop into reference plots for the guidance of control treatments, it must always be borne in mind that a plot infected by parasites is a potential source of attack on neighbouring plots. A plot in which conditions markedly favour the parasite should therefore preferably be evaluated separately, so that the findings in this plot are not "drowned" in the mean of all plots. For such parasite-prone plots, on which pests or diseases may appear earlier, then reach epidemic levels sooner, thresholds for treatments may have to be lowered. The prime consideration here is not whether the yield in this plot may be reduced, but what its infestation may mean in terms of parasite spread.

The Concept of Restricted or Selective Sampling. Since it is often neither necessary nor convenient to determine infestation in all plants or all of their parts at a given sampling point, sampling may be restricted in space or to certain organs or units. Restriction in space may consist of sampling only in locations where the pest or pathogen is expected to find favourable conditions, or where conditions are considered average, or simply where the parts to be sampled are most accessible, e.g. on the lower branches of trees. This is common practice in crop protection work. Crop parts to which sampling is commonly limited are those on which occurrence of the disease is particularly significant, e.g., tops or bottoms of branches, upper or lower surfaces or leaves, etc. The use of selective sampling to determine pesticide coverage and efficacy is discussed in Section 2.4.1.2.

Allocation of Sampling Points in the Reference Plot

Systematic vs. Randomized Sampling. The principal aim in the allocation of sampling points within the reference plot is adequate coverage. Such coverage has to ensure that no part of the crop is left without its proper effect on the result of the sampling. It can often be achieved by dividing the reference plots into a number of squares and taking samples from their centers. Alternatively, the crop may be traversed diagonally along two transects, and samples may be taken at fixed, predetermined intervals, say after every 40th step. This method, called "systematic sampling" has the advantage of being simple and can therefore be used even by untrained staff according to a plan prepared beforehand. But it may have the drawback that it cannot be used safely for statistical evaluation, which bases on randomization in order to assess the probability of error in the data obtained (Cochran 1953).

Randomized Sampling. There are several methods specifically designed for randomized sampling, and some of these will be mentioned here.

Simple Random Sampling. This consists of sampling according to coordinates which are randomly chosen anew for each observation taken in the field. This may sometimes result in large portions of the field being left unsampled, which is particularly undesirable in matters of crop protection, because early loci of pest or disease may be missed.

Stratified Random Sampling. This is a procedure in which a certain proportion of sampling units in the field is allocated at random to certain spatial "strata" which may consist of squares or other parts of the field that may be expected to be homogeneously affected, e.g., according to distance from sprinklers or from the field's border, or the strata may consist of parts of the plant expected to be infected to a uniform degree, e.g., leaves of a certain age or fruits borne at the top or at the bottom of the plant. The number of units sampled at different strata may be proportional to the size of these strata, e.g., more units may be sampled in the top 50 cm than in the bottom 30 cm of the crop; but often the number of units is more suitably fixed according to the level and variance of pest and disease occurrence in various strata.

Stratified random sampling is suitable for determination of the overall efficacy of spray application. If the aim is to determine the extent to which one particular part of the plant has been covered, such as the flag leaf of wheat (see Sect. 2.4.6.3), restricted sampling is sufficient.

Cluster sampling denotes the procedure of choosing, within the framework of a larger unit, progressively smaller units which are sampled at random. Thus, in an orchard, a sample of individual trees may first be chosen randomly. Subsequently, on each of these trees, branches are designated at random, and on each of these, random samples of leaves and/or fruits are taken.

Use of Randomized Sampling in Field and Orchard Crops. In order to introduce randomization into a systematic sampling plan, sampling points in the squares which make up the reference plots can be chosen by using tables, e.g., those published by Fisher and Yates (1963), or by aid of computer functions for generating random numbers.

In a *field crop* to be sampled at random, the sampling points within the square may be fixed as follows: the sampler stands at the center of each square and, with a random circular movement, throws a frame into the square; then he takes the sample wherever the frame happened to fall. Alternatively, when sampling proceeds along transects traversing the plot diagonally, randomization may be achieved by sampling not at fixed predetermined intervals, but at points chosen at random: in the example given above, samples would then be taken not at every 40th step, but somewhere along the line in the 40-step section. Here again, further randomization can be achieved by throwing a frame from the transect point at which the sampler stands, or by sampling highly mobile pests by sweeping with nets at random intervals. The allocation of the transects themselves may also be randomized.

Randomized sampling of *tree crops* is much more difficult, because of the pronounced ecological and developmental differences in orchards. Such differences comprise the position of the tree, whether on the edge of or well inside the orchard, or near an open space where a tree is missing; the differences within each individual tree, including its top or base, the density of its crown, the location of organs to be sampled on the inside or outside of the crown; directional aspects (N vs. S, E vs. W), and also the unavoidable lack of uniformity with which various parts of the tree are covered by spray.

Thus simple random sampling, in orchards more than in field crops, may sometimes yield misleading results, if too many samples are randomly taken at points or on sides of the tree that are less prone to attack. It is then better, as has been explained above, to resort to restricted sampling, e.g., taking samples only inside the crown or only on the western side, and to choose there only branches of a certain thickness or fruits at an age most susceptible to attack. Sampling tree crops therefore involves careful consideration of the infection pattern of the parasite for which the crop is sampled.

One of the advantages of restricted sampling is that it may reduce the cost of the operation: even though a pest may be particularly severe in tree tops, these are difficult and laborious to sample, so that in practice a smaller number of samples is made to suffice at the cost or in the time allocated for this purpose. It may then be preferable to sample solely or chiefly more readily accessible strata, but increase the number of sampling units taken there, and calibrate the results to compensate for the fact that a proportion of higher-than-average infestation is not adequately represented. However, it should not be forgotten that saving on the expenses of sampling may well become a case of "penny wise, pound foolish": a small saving in these expenses could so much increase the risk of not spraying in time or of overspraying, that the resultant loss may be much larger than the saving.

The Number and Size of Samples

The number of points at which samples are taken in each reference plot (the sample size) and the size of the sampling unit, i.e., the number of plants, leaves, fruits or other organs sampled at each point, are to some extent interdependent: in a given amount of time the sampler may examine numerous units of small size or fewer units of large size. The choice of sample size and units depends largely on the purpose of sampling, the infection pattern and severity, and on economic considerations. Relevant considerations are summarized below:

1. If infected plants are distributed in foci, i.e., if their spatial distribution is irregular, sample size should be large, while the size of the units may be small. This is a situation typically encountered in the initial stage of attack of numerous parasites, a stage of crucial importance to the plant protection adviser attempting to detect first symptoms of pathogen or pest attack.

2. If the infestation level of pests and diseases that do not typically appear in foci is low, e.g., if only a few out of thousands of leaves show symptoms, it is likewise advisable to take samples at numerous points (large sample size), especially where it is important to make sure of detecting the parasite at the earliest possible stage. If, however, the purpose is to ascertain whether infestation levels approach treatment thresholds (or whether previously severe infestation tends to decline below the threshold), then it may not be necessary to use so large a sample size: smaller samples may suffice to establish that the infestation is far below or much above the threshold. But if, in sampling for this purpose, the infestation nears the threshold level, sample size has to be increased in order to determine with greater precision how near the infestation has come to the threshold.

In stratified sampling, an optimal sampling plan will allocate to the different strata a sample size proportional to both the size of the strata and their standard deviations, and inversely proportional to the square root of their cost per unit. In the evaluation of such sampling, the results obtained in the different strata have to be calibrated to obtain a valid estimate of the average value.

3. It is often quicker and cheaper to examine a large number of plants or organs at one sampling point than fewer at more numerous points, i.e., reduce sample size and increase sampling units, although the total number of parts examined may remain constant. This, as mentioned above, is particularly true in orchards, where parasite populations have to be checked at various tree levels, and the sampler may have to climb up a ladder to reach sampling points. However, it must be borne in mind that adjacent plants or plants parts (units) are more likely to be infested similarly than plants or parts growing at some distance from one another, and this could lessen the accuracy of conclusions drawn from the sampling. In practice, therefore, a compromise must be sought to define the sample size and the number of component units at a point which involves least labour without unduly sacrificing reliability.

For the proper determination of sample size, the statistical distribution family appropriate to the pest in question has to be known from previous investigations. Moreover, the aim of sampling and the risks involved in incorrect decisions must be stated in quantitative terms. The situation obtaining in pest control can often be presented as follows. The upper threshold m_0 for preventive treatment is determined in such a way that whenever the true mean m is above m_0, the probability α will be very small that a wrong decision will be taken on the basis of the sample, e.g., that it will be decided not to spray when spraying is actually required. Similarly let the lower threshold m_1 be fixed so that the probabilitiy β will likewise be small for an erroneous decision to be taken when m is below m_1, i.e., that it will be decided to apply a treatment which is actually unnecessary. The area between m_1 and m_0 is called the zone of indifference, and as long as sampling results stay within this zone, neither decision is critical. Thus, for a given statistical distribution, after definition of the four values of m_0, m_1, α and β, the proper size of the sample n can be derived. The closer m_0 is to m_1, or the smaller the permissible probabilities of error expressed by α and β, the larger n will have to be.

To give some idea of the sample size n needed, Table 1 has been drawn up for the negative binomial distribution, with two variable parameters: m = the true mean, and k = a measure of aggregation which increases as the distribution tends to be random; for a given type of infestation and a given sampling unit size, the value of k remains fairly constant all along (Bliss and Owen 1958).

Table 1 presents the sample size n needed and the sample total S denoting the total number of infestation points which, when exceeded, indicates that control treatment is necessary; these two values are presented for probabilities of $\alpha = 0.01$ and $\beta = 0.1$, for several values of k, and for the two ratios of $m_1/m_0 = 1/2$ and $1/3$. The value $k = \infty$ (infinity) corresponds to the Poisson distribution.

To give an example from Table 1, if k is known to equal 2, and the upper and lower thresholds are $m_0 = 10$ and $m_1 = 5$ infestations per unit, then for $\alpha = 0.01$ and $\beta = 0.1$ the sample size needed is at least n = 19, and treatment will be indicated when the sum of infestation points S in a sample of 19 exceeds 120, i.e. the average per point exceeds 120/19 = 63.

Sequential Sampling

The sequential sampling technique is a method for classifying populations in which the sample size is not fixed but variable. Such sampling involves, after each observation, a choice between three possible further courses of action:

a) to continue sampling,
b) to discontinue sampling and take control action,
c) to discontinue sampling and refrain from control action.

The notable feature of this technique is its efficiency in terms of the number of observations required to reach a decision. Extremes, e.g., very high or very low infestation levels, can be established by comparatively few observations, while intermediate infestation levels will generally require more observations

Table 1. The sample size needed n (upper number) and the threshold value S for the sum of infestation points (lower number) for negative binomial distributions with different k and m values, $\alpha=0.01$, $\beta=0.1$

k/m_0:	1	2	3	4	5	6	8	10	15	20
a) $m_1=m_0/2$ (the lower threshold equals half of the upper threshold)										
1	70	51	44	40	38	37	35	34	33	32
	45	*65*	*84*	*101*	*102*	*140*	*176*	*213*	*309*	*399*
2	57	37	29	26	23	22	20	19	18	17
	37	*48*	*56*	*66*	*73*	*84*	*101*	*120*	*170*	*213*
3	52	31	25	21	19	17	16	15	13	12
	34	*40*	*49*	*54*	*61*	*65*	*81*	*86*	*123*	*151*
4	49	29	22	19	17	15	13	12	11	10
	32	*38*	*43*	*50*	*55*	*58*	*66*	*76*	*105*	*126*
6	47	26	19	16	13	13	11	10	8	8
	31	*34*	*37*	*42*	*42*	*51*	*57*	*64*	*76*	*102*
8	44	25	18	15	12	11	10	9	7	6
	29	*33*	*35*	*39*	*39*	*43*	*52*	*59*	*67*	*76*
10	44	25	18	13	12	10	9	7	6	6
	29	*33*	*35*	*34*	*39*	*39*	*47*	*45*	*58*	*77*
15	44	22	17	13	11	9	8	7	5	5
	29	*29*	*34*	*34*	*36*	*35*	*42*	*46*	*48*	*64*
∞ [a]	41	21	14	11	9	7	6	5	3	3
	27	*28*	*28*	*29*	*30*	*28*	*32*	*34*	*30*	*37*
b) $m_1=m_0/3$ (the lower threshold equals one third of the upper threshold)										
1	34	24	20	18	17	17	16	15	14	14
	17	*24*	*29*	*35*	*41*	*49*	*61*	*71*	*98*	*131*
2	29	17	14	12	11	10	9	9	8	8
	15	*17*	*21*	*24*	*27*	*29*	*35*	*44*	*57*	*76*
3	25	16	12	10	9	8	8	7	6	6
	13	*17*	*19*	*20*	*22*	*24*	*32*	*34*	*43*	*57*
4	25	15	11	9	8	7	6	6	5	5
	13	*15*	*17*	*19*	*21*	*21*	*24*	*30*	*37*	*49*
6	23	13	9	8	7	6	5	5	4	4
	12	*14*	*14*	*17*	*18*	*18*	*20*	*25*	*31*	*39*
8	23	13	9	8	7	6	5	4	4	3
	12	*14*	*14*	*17*	*18*	*19*	*21*	*21*	*31*	*30*
10	23	12	9	7	6	6	5	4	3	3
	12	*13*	*14*	*15*	*16*	*16*	*21*	*21*	*23*	*30*
15	22	12	8	7	5	5	4	4	3	2
	12	*13*	*13*	*15*	*13*	*16*	*17*	*21*	*23*	*20*
∞ [a]	21	11	7	6	5	4	3	3	2	2
	11	*12*	*11*	*13*	*14*	*13*	*13*	*18*	*18*	*26*

[a] Poisson distribution

(Wolfenbarger and Darroch 1965). Here again, the sampling plan assumes that the observations behave in accordance with a well-defined distribution, e.g., Negative Binomial, or Poisson distribution.

The decision to spray or not to spray is based on the sum of observations up to and including the one actually being made. In order to help this decision, two lines are drawn (see Fig. 1), the upper indicating the cumulative number of insects or symptoms for the number of units sampled, which definitely necessitates treatment, the lower indicating which cumulative number definitely does not require treatment. Actual sampling results are represented by a broken line as in Fig. 1. When either the upper or the lower line is traversed, sampling can be discontinued. The zone between these lines represents the zone of indifference: as long as observations stay within this zone, sampling has to continue until the full number of samples previously planned has been taken (see below).

The equations indicated on the lines in Fig. 1 depend on the distribution of the infection counts per unit sampled. Both the theory and application for Negative Binomial distribution have variously been described (e.g., Ba-Angood and Stewart 1980). With some change of notation to fit the set-up given above, the sample threshold lines are represented by:

$$S = bn + H_0 \quad \text{(lower line)}$$
$$S = bn + H_1 \quad \text{(upper line)},$$

where n is the total of sample observations taken, S is the corresponding total of infection points observed, and the slope b and intercepts H_0 and H_1 are defined

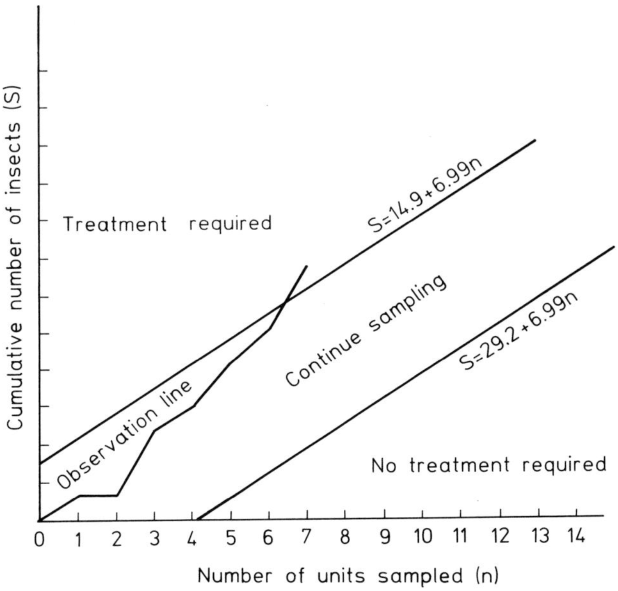

Fig. 1. Schematic representation of a sequential sampling plan. The sampling process proceeds step by step ("observation line") until a decision is reached

by the following:

$$b = k\frac{-\log C}{\log (C \cdot m_0/m_1)}$$

$$H_0 = \frac{\log \alpha - \log (1-\beta)}{\log (C \cdot m_0/m_1)}$$

$$H_1 = \frac{\log (1-\alpha) - \log \beta}{\log (C \cdot m_0/m_1)},$$

where $C = (m_1 + k)/(m_0 + k)$ (for definition of k, m_0 and m_1 see section on number and size of samples, p. 63).

The lines in Fig. 1 are actually drawn to scale in the set-up of the example given above: the values were found to be $b = 6.99$, $H_0 = 29.2$, and $H_1 = 14.9$.

For the Poisson distribution, H_0 and H_1 remain unchanged with $C = 1$, and the slope b becomes $b = (m_0 - m_1)/\log (m_0/m_1)$. For the normal distribution the corresponding expressions are $b = (m_0 + m_1)/2$, and $\log(C \cdot m_0/m_1)$ and H_0 and H_1 change into $(m_0 - m_1)/\sigma^2$.

To plot the line of observations, beginning at 0 and progressing between the threshold lines until one of the latter is traversed, proceed as follows: for every unit inspected move one space to the right, and if the unit is infected also move upwards by the number of infestation points (e.g., insects) observed. If there are few infected units, the observation line will fairly soon traverse the lower threshold line and sampling is discontinued, because treatment is clearly not necessary. But if there are numerous infected units, the observation line will traverse the upper threshold line and treatment is indicated. Thus, the sooner the observation line traverses either threshold line, the greater is the saving in sample size as compared to a fixed sampling plan.

Successive Sampling

Where the sampling procedures described so far are used, the decision to treat or to delay treatment rests solely on the results of the sampling just carried out. This has several disadvantages:

a) Since previous assessments are not necessarily taken into account, the dynamics of pest and disease development is not calculated. This may influence the decision for control, e.g., where infestation levels hover for some time just below treatment thresholds.
b) Errors in one-time sampling are harder to detect, whereas a succession of samplings, the results of which are related to each other, affords a good chance to detect discrepancies due to error.

Weighting or Calibration

In restricted sampling, units are chosen for their convenience or for their significance in denoting infestation or out of other practical considerations. Also, as noted above, in stratified sampling the number of sampling units may purposely be disproportional to the size of the strata. Simple means of the result of such sampling will thus not, in general, reflect the true overall infestation level in the reference plot. In order to determine this level from samples taken by the above procedures, sampling results have to be calibrated or weighted: each observation has to be multiplied by a certain factor to correct the disproportionate representation of units. This factor has to be fixed a priori when the nature of the unit to be sampled (e.g., the most susceptible leaf or most harmful size of larva), or the allocation of that unit (e.g., at the top of the slope) is defined before the sampling begins.

A slightly different reason for calibration is found in the evaluation of samples taken in monitoring *Heliothis* in cotton in Israel (see Sect. 2.4.6.2). Here the larvae found at each sampling point are calibrated according to their damage potential (life expectancy and feeding capacity): larvae measuring up to 15 mm in length are given a calibration factor of 2, while larger larvae are given only a factor of 1.

Drawing Up a Plan for Threshold Sampling

In accordance with all that has been said above, a number of steps have to be taken to draw up a sampling plan best suited for the purpose:

a) Choose an appropriate sampling procedure, e.g., stratified sampling, and sampling units, such as three consecutive plants, or the lower sides of leaves.
b) Define quantitatively the threshold for treatment, i.e., the mean infestation level of your reference plot that warrants treatment, and the size of the zone of indifference below this threshold where either decision (to treat or not to treat) will be acceptable.
c) Try to find, either in literature or on the basis of previous sampling, the statistical distribution (e.g., Poisson or Negative Binomial) most suitable for observations on the pest or disease which is the object of your sampling.
d) Establish, by use of tables appropriate for predetermined rates of error α and β, the size of the sample required for a fixed sampling plan, or the boundaries for decision in a sequential sampling plan.
e) Construct the actual sampling plan, including the sampling points.

References

Ba-Angood SA, Stewart RK (1980) Sequential sampling for cereal aphids on barley in Southwestern Quebec. J Econ Entomol 73:679–681

Bliss CI, Owen ARG (1958) Negative binomial distribution with a common K. Biometrika 45:37–58

Cochran WG (1953) Sampling techniques. Wiley, New York

Fisher RA, Yates F (1963) Statistical tables. Oliver & Boyd, London

Ruesink WG (1980) Introduction to sampling theory. In: Kogan M, Herzog DC (eds) Sampling methods in soybean entomology. Springer, Berlin Heidelberg New York

Wolfenbarger DA, Darroch JG (1965) A sequential sampling plan for determing the status of corn earworm control in sweet corn. J Econ Entomol 58:651–654

2.4.1.2 Selective Sampling for Monitoring Application Efficacy

H. FRANKEL

Sampling methods and statistics used to evaluate pesticide application techniques and their field efficacy must be specially adapted to this purpose. Two main criteria have to be considered to this end:

a) The distribution variability obtaining with the machine or technique used (see Chap. 2.5.5) and the collection and retention efficiency of the target (see Chap. 2.5.3).
b) The sum total of all shortcomings in application, e.g., differing dosage levels and leaf surface coverage lacking uniformity as well as the uneven spread of heterogeneous pest populations, and the different dose-response levels of the latter, are factors that greatly influence control efficacy.

Variability of Deposit. The range of variability in cover deposit which is obtained when sampling for the determination of application efficacy, depends in part on the machine, the technique used, and the characteristics of the crop. Variability may be slight in placement spraying of low crops, when these are treated with automated spray booms, but may be somewhat greater when such crops are treated with hand-operated equipment. Other field crops, and especially "space crops", such as irrigated cotton or maize, or plantation crops, are much harder to cover evenly on all surfaces by any type of equipment. Very extensive variability occurs in application of sprays by aircraft to space crops, and this is exceeded only by the variability of deposit resulting from drift spraying.

It follows that sampling for uniformity of coverage cannot be randomized without restriction, as is often the practice in threshold sampling, but should be selective. Such sampling aims at the evaluation of samples taken at predetermined points located at given distances across treatment swaths. Thus sampling of a spray boom is made across its swath on artificial targets, such as glass or mylar for volumetric recovery or paper strips for qualitative assessment. The distance of targets from one another is a function of the width of swath of a single nozzle and its overlap to neighboring nozzles. At least three replications should be made, and the data are to be presented as maximum, mean, and minimum values for each sample point tested. For aircraft or air-assisted spraying, at least three swaths should be sampled to measure the effect of the overlap, and three to five replications should be made. Variability determined in this way is most conveniently expressed as a percentage of the mean, i.e., the coefficient of variation. This indicates the sample variability above and below the mean with an overall probability of 68%.

It should be noted that variability is in part a function of the dimension of the target sampled; it will frequently decrease in inverse proportion to target size, i.e., will be smaller on large than on small leaves. The size of the sample area must therefore be constant, and in order to be meaningful, it must be relevant to the actual size and shape of the target and to the cover density required to control a given pest on its target site. It follows that in testing machines or techniques by sampling artificial targets, the values obtained must be considered indicative only. To obtain more truly representative results, tracer techniques can be used, which visually indicate deposit quality on the crop itself (see Chap. 2.5.7); most efficient of all is the combination of bio-assay tests with chemical analysis of the recovered residue.

Control Efficacy. Sampling to assess treatment efficacy in commercial crops is often complex, especially when "space crops" have been treated with reduced volumes, by air-assisted or drift techniques, or by aircraft. In such crops the wealth of multiplane surfaces stacked in various foliage layers causes extreme variance in the deposits retained. If customary threshold count techniques were to be used for sampling such crops, this would entail taking hundreds of samples to obtain truly representative values with a high order of probability. This is simply not practical, and selective sampling must therefore be employed. If, for example, previous counts have shown infested areas to have three levels of pest incidence, randomized subplots should be selected and marked for volumetric recovery, for bio-assay, or at least for qualitative analysis: whilst the first method will indicate the amount of deposit, the second will show pest response, and the last, in conjunction with field counts, will suffice to indicate deposit values as related to control efficacy.

To give an example: the variability of deposition of spray released from an aircraft has been shown to be $\pm 80\%$ around the mean, as measured by volumetric recovery on individual leaves at the top of the plant (Chap. 2.5.5). But at the crop's bottom layer, the amount of deposit recovered is only 25% of that recovered at the top. Thus, if the mean value at the top is $2\ \mu g\ cm^{-2}$, the leaves receiving the lower range of deposit at the top (-80%) received only $0.4\ \mu g\ cm^{-2}$ and those in the bottom layer received $0.1\ \mu g\ cm^{-2}$. The differences in control efficacy resulting from such distribution are obvious, and hence selected layer sampling is imperative to overcome application variability.

In monitoring application efficacy, the determination of mean values for the crop is thus clearly irrelevant, and the variability rather than the mean of the deposit is what matters. Most important may be those plant parts that have not been adequately covered: these are potential sources of continued epidemic or endemic pest and disease development. After pesticide application to vegetation, the accent in sampling must thus be placed on those host sites that are least accessible to treatment. Moreover, sampling methods must take due account of the characteristics of pest behaviour.

2.4.2 Monitoring and Prognosis by Environmental and Empirical Criteria

J. PALTI and R. AUSHER

When is the pest or disease due to appear, and once it has appeared, how severe is the attack? For a fair number of economically important parasites schemes have been worked out, most in the developed countries, to predict either the period in which they will *not* appear (negative prognosis), or the combination of environmental and biological factors that will lead to their appearance (prognosis, forecasting). But for a large number of parasites in the majority of countries, such predictive schemes are not available. In these cases, the likelihood of parasite appearance can be defined by the establishment of empirical threshold dates, and their actual appearance by trapping of insects (including virus vectors), mites, and sometimes spores, and by monitoring or field scouting.

Negative prognosis and forecasting on a regional scale is possible only where (a) regional weather parameters can be measured and processed efficiently, and (b) crop conditions are reasonably uniform in all that relates to plant development (growth stage, density) and environmental factors (leaf wetness, irrigation), or in cases in which such factors can be recorded in the fields of individual growers, as practiced in the United States in forecasting potato blight by the Blitecast method (see Sect. 2.4.2.3).

The following description of means of predicting or monitoring parasite attacks on crops is limited (with some exceptions) to practical means available to a large number of farm advisers. But reference will also be made to regional or country-wide monitoring schemes developed more recently by aid of electronic equipment, and made available to the adviser and grower (see Sect. 2.4.6.1).

In some east European countries a distinction is made between (a) pest and pathogen monitoring, which is general and centralized, run by the state, and uses long-term data bases for predicting the likelihood of development of pest or pathogen populations that may cause economic damage; and, based on this, (b) field-specific monitoring carried out by crop producers, which relates to treatment thresholds that take into account pest, crop and weather conditions, and aims at actual, operative pest control decisions (Wetzel et al. 1982).

2.4.2.1 Pre-Season Weather and Infection Criteria for Prognosis and Monitoring

Weather and Level of Infection or Infestation in Preceding Seasons. The infection level in the preceding crop season and/or the weather, especially the temperature, in intervening seasons, are of special significance for the timing of monitoring activities in the following cases: (a) in perennial crops, more particularly in deciduous orchards; (b) in crops propagated by vegetative parts, such as bulbs, corms, set tubers, etc.; (c) in crops subject to attack by pests whose overwintering stages are susceptible to the cold, or to attack by diseases whose vectors are similarly susceptible.

A well-known example of the latter case is that of the bacterial wilt of corn (maize) (*Xanthomonas stewartii*), predicted according to temperatures of the preceding winter: if the sum of the mean monthly temperature for December, January and February equals less than 32 °C, this indicates practical absence of the disease during the following season, while a sum of 32–38 °C or more indicates danger of epidemic development. This "winter temperature index" is known to be related to the survival of the wilt's vector, the corn flea beetle (*Chaetocnema pulicaria*), in whose alimentary canal the causal bacterium overwinters (Anonymous 1968).

It has also been shown that the time of appearance of tobacco blue mold (*Peronospora tabacina*) in the United States is related to the temperature of the preceding January. If this is above normal, the mold appears in seedbeds early in the growing season and causes heavy damage, but when January temperatures are low, the disease appears late and causes little damage (Anonymous 1968).

As regards survival of a pathogen in crop organs and its dependance on weather, Jeger and Butt (1983) found that severe winter weather (temperatures below −12 °C) will kill the mycelium of apple powdery mildew (*Podosphaera leucotricha*) in most apple buds and thus reduces the level of disease in spring, which often obviates spraying early in the season. Similarly, in countries in which grapes are attacked by downy mildew (*Plasmopara viticola*), but the fungus does not form overwintering oospores, the mildew in mild winters often persists on leaves that have not shed and then attacks the next crop early in spring; but if the winter is colder, no such leaves persist on the vines, and mildew attacks are delayed.

In the U.S.S.R., the influence of the preceding season's infestation level and of winter weather has been studied in relation to many pests. Thus forecasts for the purpose of monitoring and control are made for cockchafers (*Melolontha* spp.) and true wireworms (Elateridae) according to their distribution and abundance in the previous year. Forecasts of the cutworm *Agrotis segetum* take into account viability of the population in the previous season, as well as temperatures from autumn to spring: in cold winters the overwintering larvae penetrate into the soil and take some time to pupate and hatch a late spring generation, but in mild winters they continue to inhabit upper soil layers and to feed, and then produce an early spring generation. Population viability in autumn also serves for preliminary forecasts of the European Corn Borer (*Ostrinia nubilalis*), but such forecasts are subject to change, depending on the weather in spring.

Another way in which winter weather affects pest survival is through its effect on plants that carry the pest over the winter. Turl (1983) has reported that the extent to which the aphid *Macrosiphum euphorbiae* develops on potato crops in Scotland depends largely on the cold-hardiness of weeds that harbour the aphid in the preceding winter. Since the preferred host of this aphid is the weed *Urtica urens* L., and this is not hardy, lighter and later aphid attacks on potatoes may be expected after cold than after mild winters.

As regards pre-season criteria for forecasting diseases, two examples will be given, both relating to the carry-over of inoculum from one season to the next in east European countries.

In Poland, Rondomanski (1969) has shown that the extent to which onion propagating material is infected by downy mildew (*Peronospora destructor*), can be used as an indication how severe the disease may be expected to be in the forthcoming season. This depends on the percentage of infected mother bulbs, which in turn is determined by the previous season's weather: if temperature and humidity conditions were then conducive to rapid dessication of foliage, the fungus is taken to have penetrated only few bulbs (less than 0.1%) with consequently low systemic infection. But if the growing season has been prolonged, more bulbs have been infected (1% or more) and are transmitting the mildew to the next crop.

The PHYTEB prognosis scheme developed in the German Democratic Republic (Gutsche and Kluge 1983), uses the amount of potatoes affected in the previous autumn's yield to estimate the length of the period over which the spring potato crop is expected to remain free from *Phytophthora* blight (see Sect. 2.4.6.1).

Winter Monitoring of Pests on Branches of Deciduous Fruit Trees. In deciduous orchards in temperate climates, it is important to determine the quality and size of the overwintering pest population, in order to verify two aspects: whether pests are present that may require treatment before the start of the growing season, and to gain an idea which of the pests harmful during the growing season will need watching, how large their overwintering population is, and to what extent it is parasitized. For this purpose a method has been developed in central Europe (IOBC 1980) which consists of checking by magnifying glass or binocular ten branch sections of 20 cm length (five samples per plot) to total 2 m per sample. The samples are taken from 2–3-year-old wood well provided with spurs and laterals. A few such examples are listed in Table 1.

Table 1. Treatment thresholds for apple pests as ascertained by branch sampling of overwintering stages (IOBC 1980)

Pest	Treatment threshold (total for 10×20 cm branch samples)	
Operophthera brumata (winter moth)	2–5	eggs
Hyponomeuta malinellus (small ermine moth)	2–5	hatchings of small larvae
Aphis pomi (green apple aphid)	5–50	eggs
Psylla mali (apple sucker)	200–500	eggs
Panonychus ulmi	1000–2000	eggs

2.4.2.2 Empirical Threshold Dates of Pest and Disease Appearance

Ascertaining whether or not a disease or pest has appeared is expensive, especially if it involves the operation of monitors or field scouts. Limiting the period in which such monitoring has to be carried out can therefore save considerable outlay. Two means to this end are the establishment of threshold dates for pest and disease appearance, which can also be described as the definition of pest- or disease-free periods, and the trapping of insects or spores.

Establishment of threshold dates is of considerable practical value wherever consecutive seasons differ markedly in weather, e.g., alternation of warm and cold or of dry and rainy seasons. Systematic observations, carried out over a number of years (the more, the better) help to establish purely empirical dates *before* which a given pest will not appear in the region, no matter what the cultural conditions under which the crop is grown (date of sowing, density, neighborhood, etc.). Once such a disease- or pest-free period has been determined, its end can be considered as the threshold date of appearance of the parasite in question, and monitoring can be directed to cover only the period following the threshold date.

Keeping records for this purpose, establishing card indices and maintaining them over the years is the simplest method of collecting data on pest and disease appearance that are relevant to local cropping systems, e.g., to fields sown at various densities or on various dates, to crops grown with or without irrigation, etc. It is most important that a continuous check be kept on the reliability of observations supplied for recording, especially where these seem out of the ordinary, and to make sure that no observations of doubtful validity ever get into the card index.

Observations carried out by regional plant protection advisers and recorded in such card indices, have made it possible to determine, in Israel, country-wide thresholds for the appearance of about 25 diseases and pests of annual crops, and thus to lighten the adviser's and the farmer's burden of monitoring. Some examples of such threshold dates, for spring and autumn vegetable and field crops are presented in Table 2. Threshold dates for perennial crops are less useful, because the extent to which the parasite succeeds in passing unfavourable seasons is more difficult to determine accurately.

2.4.2.3 In-Season Monitoring According to Temperature and Humidity Effects on Pest and Disease Appearance

A common aid to the proper timing of pest and disease monitoring and control, other than trapping of parasites, is the determination of environmental factors that advance or delay appearance of pests or pathogens before or during the season of crop growth. This can sometimes be done on the basis of one factor alone, mostly temperature, but more often takes into account humidity factors as well as temperatures. Examples for this kind of prediction are so numerous that we shall here confine ourselves to a few cases and to the citation of references for further study of the subject.

Temperature Sums or Degree Days. The idea of predicting first seasonal appearance or generation time, mostly of pests, by summing up temperatures sup-

Table 2. Country- wide empirical threshold dates for parasite appearance on some outdoor vegetable and field crops in Israel

Crop	Month of sowing	Disease or pest	Threshold date	Remarks
Melon	March	Powdery mildew (*Sphaerotheca fuliginea*)	1st week May	Earlier in close vicinity to crops under plastic cover
Carrot	February	Powdery mildew (*Oidium* sp.)	1st week May	
Melon	March	Red melon beetle (*Rhaphidopalpa foveicollis*)	1st week April	
Tomato	March	Powdery mildew (*Leveillula taurica*)	2nd week May	
Onion	October	Onion fly (*Hylemyia antiqua*)	1st week Nov.	
		Downy mildew (*Peronospora destructor*)	1st–2nd week January	Earlier in seedbeds or on volunteer plants
Cucumber	March	Downy mildew (*Pseudoperonospora cubensis*)	1st week May	Earlier in close vicinity to crops under plastic cover
Broad bean (*Vicia faba*)	October–November	Rust (*Uromyces fabae*)	2nd week Dec.	
		Chocolate spot (*Botrytis fabae*)	1st week Dec.	
Lucerne	Autumn	Alfalfa weevil (*Hypera variabilis*)	1st week Nov.	
Groundnut	April	Leaf spot (*Cercosporidium personatum*)	4 th week June	Earlier on groundnuts sown after groundnuts
Sugarbeet	September–October	Beet weevil (*Lixus junci*)	2nd–3rd week Dec.	
		Powdery mildew (*Erysiphe betae*)	3rd week Dec.	
		Rust (*Uromyces betae*)	2nd week Jan.	

porting their development is attractive, but in practice such predictions are less simple than they may appear at first glance. They involve familiarity with the pest's biology: determination of the minimum temperature for its development, and often also of the maximum temperature, because only temperatures within this range are fully relevant. Thus, in California, the minimum temperature for the development of the cotton bollworm (*Heliothis zea*) is a daily mean of about 12 °C, a threshold below which temperatures do not commonly fall in the cotton-growing season. However, the maximum daily mean is 34 °C, a threshold occasionally exceeded and potentially harmful to the pest, and this temperature could render the value of the temperature sum on certain days somewhat dubious.

Another point to consider is at what time to begin the process of recording temperatures for the purpose of calculating biologically meaningful temperature sums. A few examples from California will illustrate this point. There are basically two options: to record and sum up the temperatures from a set date onward, or to begin the process when the first generation makes its appearance. Set dates are used as points of departure for a variety of pests. Thus, temperature sums for the Lygus bug (*Lygus hesperus*) affecting cotton and safflower are calculated from April 1st onward (Wilson and Barnett 1983), that for the pink bollworm (*Pectinophora gossypiella*) on cotton from January 1st (Sevacherian 1978) and that for the San Jose Scale (*Quadraspidiotus perniciosus*) in orchards from early March (Rice et al. 1982). For other pests, however, the starting date for recording temperatures (zero time) may be the first day on which a female adult is caught, as is the practice with regard to the California Red Scale (*Aonidiella aurantii*). Table 3 lists the developmental thresholds, in degree-days per generation, of nine common pests of crops in California.

The significance of degree-day values may vary with the pest's biology in various regions. Thus the alfalfa weevil (*Hypera brunneipennis*) in some regions of Illinois (U.S.A.) lays eggs in autumn and winter, and there accumulation of 200 degree-days (above 48 °F, calculated from January 1st) indicates the time when sampling has to begin; however, in regions in which the weevil does not lay eggs in these seasons, sampling can be delayed until 400 degree-days have accumulated (Wedberg et al. 1980).

It must be stressed that temperature sums or degree-days calculated in the above ways sometimes represent only approximate values, since day length, humidity conditions, and the range of seasonal temperature fluctuations affect the biological effectiveness of temperature levels in many cases (Makarova et al. 1981). However, in spite of the limitations mentioned, the calculation and use of heat units, in the form of degree-days, is a useful tool for pest management, as aptly described by Zalom et al. (1983). It can tell growers and advisers where they stand in relation to the development of a generation of pests, so that they can time pesticide applications more efficiently. For example, pheromone traps might indicate an increase in the number of adults of a pest species. The accumulated degree-days would show whether this is a real or a false peak, and whether treatment should be promptly applied or can be delayed. Degree-days can also be used to guide sampling and to co-ordinate cultural and chemical practices, e.g., the times of irrigation and pesticide treatment.

Table 3. Developmental thresholds °C and physiological generation times for selected pests. (By courtesy of Dr. Frank Zalom)

Organism	Developmental thresholds (°C)		Degree-days[a] per generation	Starting date for recording temperatures	Crop(s)
	Lower	Upper			
California Red Scale *Aonidiella aurantii*	11.6	35.0	572	Trap first flight April–May, then start recording temperature	Citrus
Codling moth *Laspeyresia pomonella*	11.0	34.4	603	Biofix determined once there have been two twilight temperatures about 16.6 °C along with male codling moth trapping	Walnuts
Cotton bollworm *Heliothis zea*	13.8	34.4	538		Cotton
Egyptian alfalfa weevil *Hypera brunneipennis*	7.2	–[b]	444		Alfalfa
Lygus bug, *Lygus hespersus, L. elisus*	11.0	–[b]	365	April 1 in safflower	Cotton
Oriental fruit moth *Grapholitha molesta*	7.2	32.0	535	February 15.	Peaches, nectarines, quince and almonds
San Jose Scale *Quadraspidiotus perniciosus*	10.5	32.0	583	Early March, usually control applied with winter dormant spray	Walnuts
Walnut aphid *Chromaphis juglandicola*	5.2	–[b]	318	May	Walnuts
Pink bollworm *Pectinophora gossypiella*	15.5	–[b]	537	January 1	Cotton

[a] Degree-days provide a valuable tool for managing insect pests. It is most useful in predicting pest occurrence, scheduling and initiating pest control actions, and monitoring pest activity

[b] No upper threshold limits available

Prediction by Combined Temperature and Humidity Factors. Although temperature alone can serve as a guide for predicting the appearance of certain pests, the humidity factor enters into almost all the schemes evolved for prediction of crop diseases. Humidity may variously be expressed as the macroclimatical factors of rains or fog, or as the microclimatical factors of dew periods or "wetness periods" due primarily to dew but also to rain, fog, or overhead irrigation. The biological effect of humidity is, with few exceptions, related to prevalent temperatures, and often also to the stage of crop growth and/or pathogen factors.

The subject of monitoring diseases has recently been reviewed by Young et al. (1978), though this review refers to large-scale, regional monitoring for survey purposes rather than to the monitoring of individual fields with which we are chiefly concerned.

With very few exceptions, prediction schemes based essentially on temperature and humidity factors have as their aim the prediction of diseases that affect crops under the conditions of temperate regions, mostly unirrigated crops. Under such conditions, the temperature and humidity regime of many fields of a crop in a given region can be uniform enough to facilitate *regional* forecasts applicable to all or most of the fields with that crop. But in relation to irrigated crops this is rarely feasible: each crop and its pathogen develop differently, according to the time, frequency, and rate of irrigation, so that regional forecasts may be of little value. However, in recent years, *field-specific* forecasts have become possible by the introduction of systems such as *Blitecast* (see below), where recording equipment placed by the farmer in his field is used to report to a central computer which then issues a warning applicable only to that field, if conditions there favour disease outbreak. If similar equipment becomes available in tropical and subtropical regions, it could probably be used to forecast many more diseases that cause losses in these regions.

The following examples for guidance of disease prediction, monitoring and control on the basis of humidity and temperature parameters have been selected to indicate the great variety of ways in which this can be done.

Downy Mildew of Grapevine (Plasmopara viticola). This downy mildew was the first to be forecast, on the basis of rules worked out between 1912 and 1922 in France and Germany. The various systems now in use have been reviewed by Kable (1978). They are primarily based on the assumption that the mildew will cause first infections when (a) daily minimum temperature exceeds 10 °C for 2–3 days; and (b) there has been rainfall of 10 mm or more within this period.

In many regions adaptations of this rule have been found useful for more precise predictions, e.g., minimum temperatures of 12 °C; the growth stage of the vine is also often used as an auxiliary criterion, e.g., shoots must be at least 10 cm long, or oldest leaves at least 2.5 cm in diameter. It has also been found possible in some European vine-growing regions to fix threshold dates for downy mildew appearance, mostly between 1st and 15th May. The presence of inoculum is taken for granted.

The schemes for downy mildew prediction in vineyards are thus regional schemes of positive forecasting, aided by in-field criteria such as growth stage (see Sect. 2.4.2.4).

Potato Late Blight (Phytophthora infestans). Two schemes for late blight monitoring and forecasting will be described because they differ so widely from one another.

The PHYTPROG system operated in the Federal Republic of Germany (Ulrich and Schroedter 1966) is designed to assess disease-free periods on the basis of temperature and humidity, by calculations on the following basis:

a) Hours in which relative humidity is above 90% or precipitation at least 0.1 mm (only groups of not less than 4 consecutive hours are taken into account).
b) Hours with temperatures between 10° and 24 °C (at least 4 consecutive hours).
c) Infection factors worked out for each temperature level, e.g., for 10°–12 °C, the infection factor is 0.8990, for 16°–18 °C the infection factor is 0.536, etc.

When the sum of the product of hours with the requisite humidity and temperature, multiplied by the appropriate infection factors, reaches a threshold of 150, monitoring is advised although the disease hardly ever appears before this sum reaches 270. The calculations begin on the day the potato sprouts emerge. Provisions are made to take into account hours of low humidity which slow down blight development.

This computer-based system has been working well in Germany ever since 1967. It encounters difficulties where potato crops have their humidity conditions altered by irrigation. It is a scheme for negative regional blight prediction.

The BLITECAST for predicting outbreaks of the same disease, as described recently by Mackenzie (1981), has two triggers, either of which can issue a recommendation to begin monitoring or consider spraying. One trigger is pulled after 10 consecutive days when the cumulative rainfall exceeds 30 mm and the 5-day average daily temperature does not exceed 25.5 °C. The other trigger is pulled when the accumulated "severity values" exceed 18–20 units. These units, which have been given arbitrary values of 0 to 4, are assigned for each day according to the extent to which its temperature and humidity levels favour the blight: a very unfavourable day, e.g., with temperatures below 10 °C and only 2 h of relative humidity greater than 90%, would be given a value of 0, while a day with 19 h relative humidity above 90% and an average temperature of 21 °C would be given a value of 4. The accumulation of weather data begins when rows of emerged plants are first visible in the field. The data can be collected by battery-operated microcomputer which is placed in the field and also interprets the data recorded, i.e., "pulls" the appropriate trigger. From the aspect of monitoring disease appearance, this system serves for positive forecasting of late blight in the field of each grower equipped with the required microcomputer.

In both the systems described above, the presence of inoculum of *Phytophthora infestans* is taken for granted. This assumption has in most cases proved right for the regions in which the systems are used, i.e., the temperate regions in which potatoes are widely grown. However, a similar assumption cannot be made in other climates, e.g., in semi-arid regions, where predictions without verification of the presence of inoculum are too risky to be of practical value.

Thus, the PHYTEB blight prognosis (Gutsche and Kluge 1983), as already mentioned in Section 2.4.2.1 and described more fully in Section 2.4.6.1, has been developed in the continental climate of the German Democratic Republic and uses "infection pressure" in its prediction scheme. The size of infection pressure is estimated according to the incidence of potato blight in the previous season's crop, and provides an indication of the period expected to be free from disease.

2.4.2.4 The Growth Stage or Age of the Crop as Guide for Monitoring

The stage of growth reached by the crop is the criterion most convenient to link with prospective parasite appearance, wherever this can be done reliably. In some cases the growth stage alone can serve as such a criterion, in other cases it must be viewed in conjunction with weather and/or parasite factors.

Useful classifications of growth stages of field crops, viz. wheat, barley, oats, rye, maize, rice, tobacco, cotton, sunflower, potato, broad bean (*Vicia faba*), rape and soybean have been published in the crop loss manual of the FAO (Chiarappa 1973). These classifications are more useful to the adviser than more elaborate scales of growth stages published for research purposes. Some of the FAO scales are reproduced in Figs. 1–3 with indications at what particular stage certain pests should begin to be monitored in various countries.

Figs. 1 and **2** see pp. 80 and 81

Fig. 1. Growth stages in wheat, oats, barley, rye (Chiarappa 1973). The growth stages of wheat and barley (G.S.) at which some pests have been found to make their first appearance in various countries have been listed by Benedek et al. (1981) as follows:

Eurygaster integriceps	– at G.S. 0–4 in U.S.S.R.
Zabrus tenebrioides	– at G.S. 5–10 in U.S.S.R.; at G.S. 10.5–11.1 in the German
Oulema spp.	Democratic Republic
Oulema spp.	– at G.S. 5–10.1 in U.S.S.R.
Oscinella frit	– G.S. 10–10.5 in the German Democratic Republic
Haplodiplosis equestris	– at G.S. 0–4 in U.S.S.R.
Sitodiplosis mosellana	– at G.S. 9 in Czechoslovakia
Contarinia tritici	– at G.S. 10.5 in Czechoslovakia
	– at G.S. 10.5 in Czechoslovakia

Substages in stages 10 and 11: *10.1* first ears just visible; *10.2–10.5* progressive stages of the heading process; *11.1–11.4* progressive stages of kernel development, from milky ripe to ripe for cutting

Fig. 2. Growth stages of maize (Chiarappa 1973). The growth stages (G.S.) at which some maize pests have been found to make their first appearance in various countries:

Popillia japonica	– at G.S. 8 in Indiana, U.S.A. (Chiarappa 1973)	
Busseola fusca *Sesamia calamistis*	– stem borers, at G.S. 2–3 in Tanzania (Chiarappa 1973)	
Rhopalosiphum maydis	– aphid, at G.S. 5–6 (3 weeks before tasseling) in Indiana, U.S.A. (Chiarappa 1973)	
Tanymecus dilaticollis *Anoxia villosa* *Agrotis segetum*	– southern gray weevil – steppe cockchafer – cutworm	in Romania, several days *before* planting (Chiarappa 1973)
Ostrinia nubilalis	– at G.S. 4–6 in U.S.S.R. (Benedek et al. 1981)	
Oscinella frit	– at G.S. 1 in U.S.S.R. (Benedek et al. 1981)	

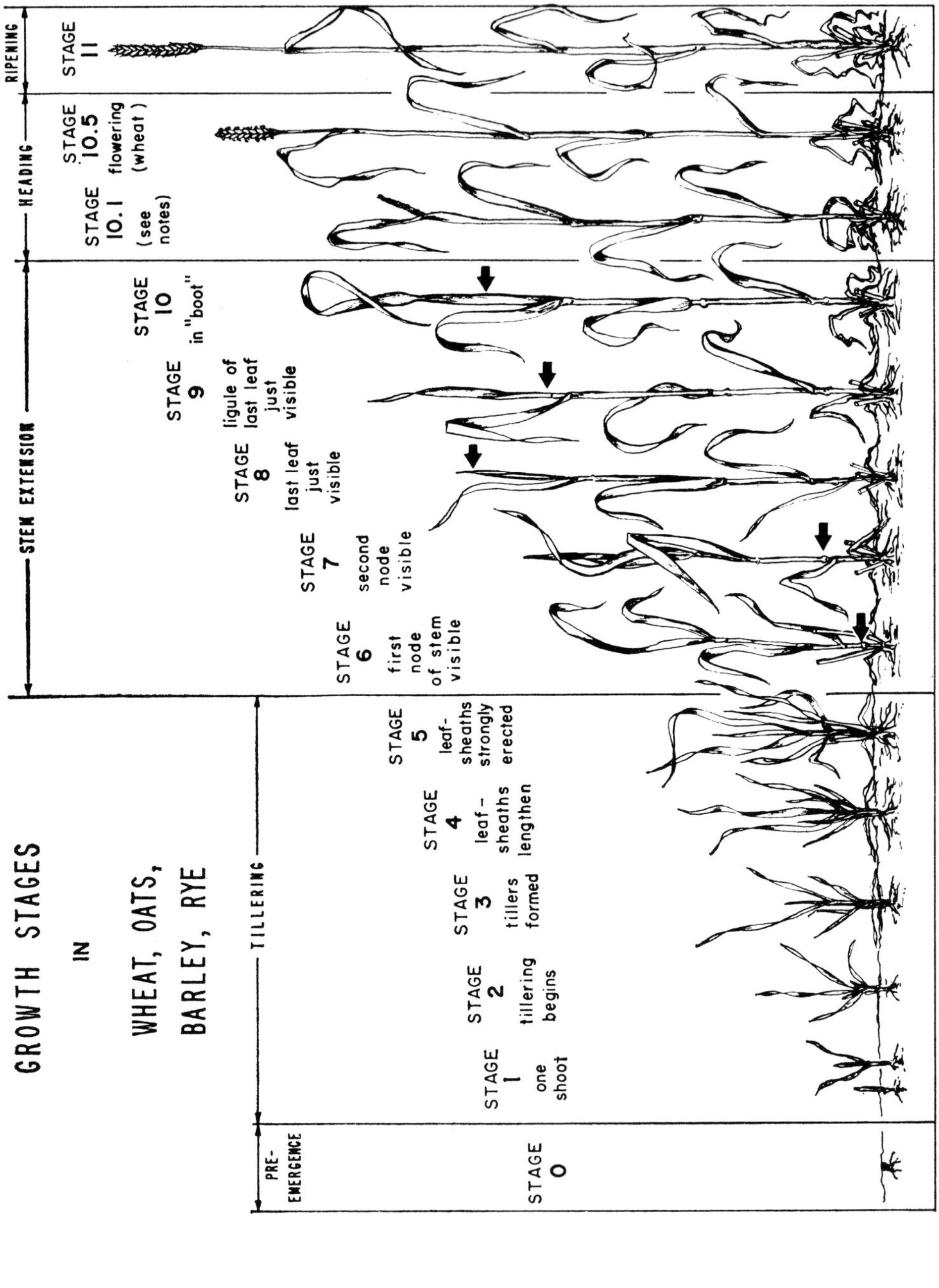

Fig. 1 (legend see p. 79)

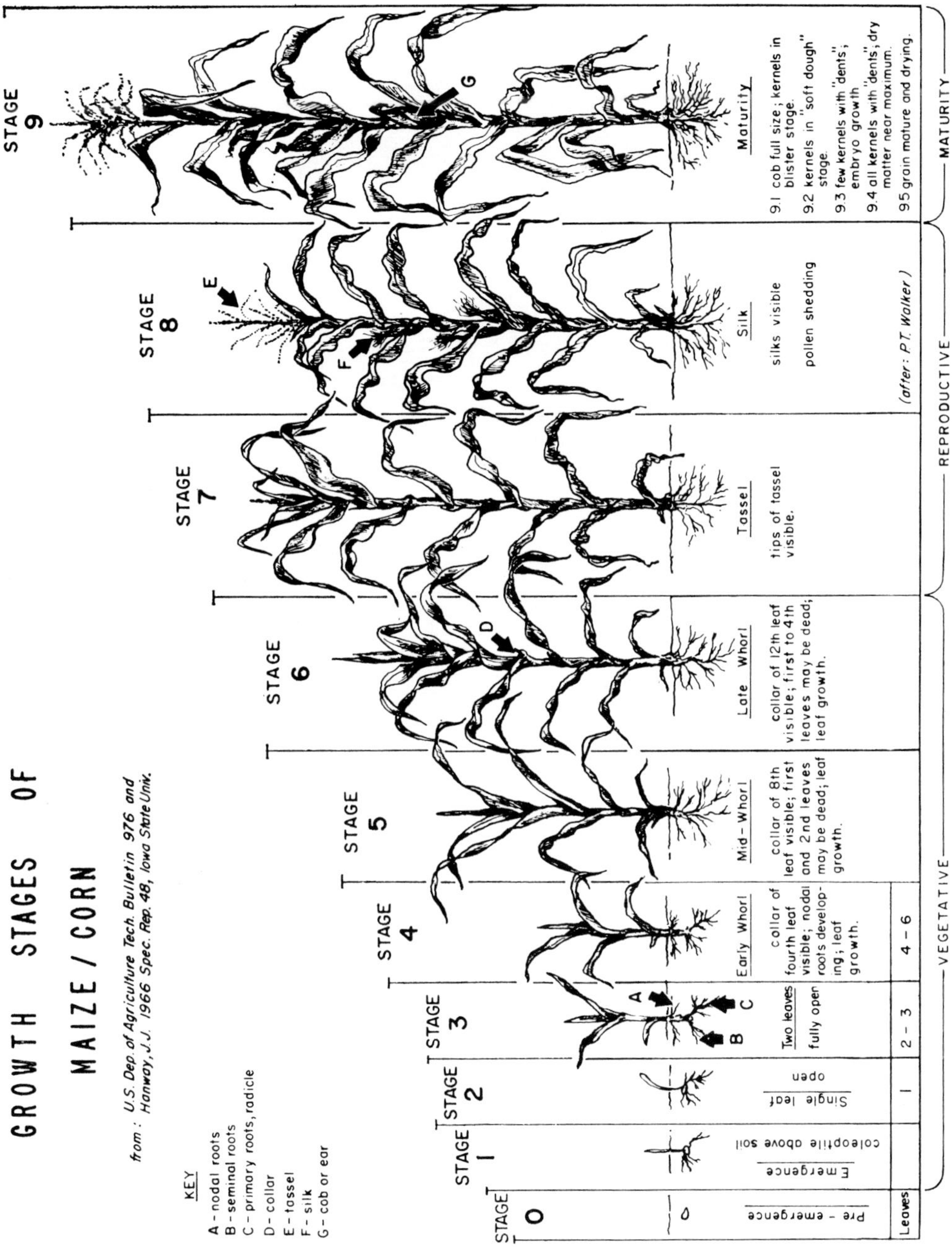

Fig. 2 (legend see p. 79)

Fig. 3. Growth stages of rice (Chiarappa 1973). The growth stages (G.S.) at which some rice pests have been found to make their first appearance have been listed by Chiarappa (1973) as follows:

Niliparvata lugens – brown leafhopper, at G.S. 3 (end of tillering) in Japan
Oebalus pugnax – rice stinkbug, at G.S. 5 in Texas, USA
Sogatella furcifera – white-backed planthopper, at G.D. 5–6 in Japan

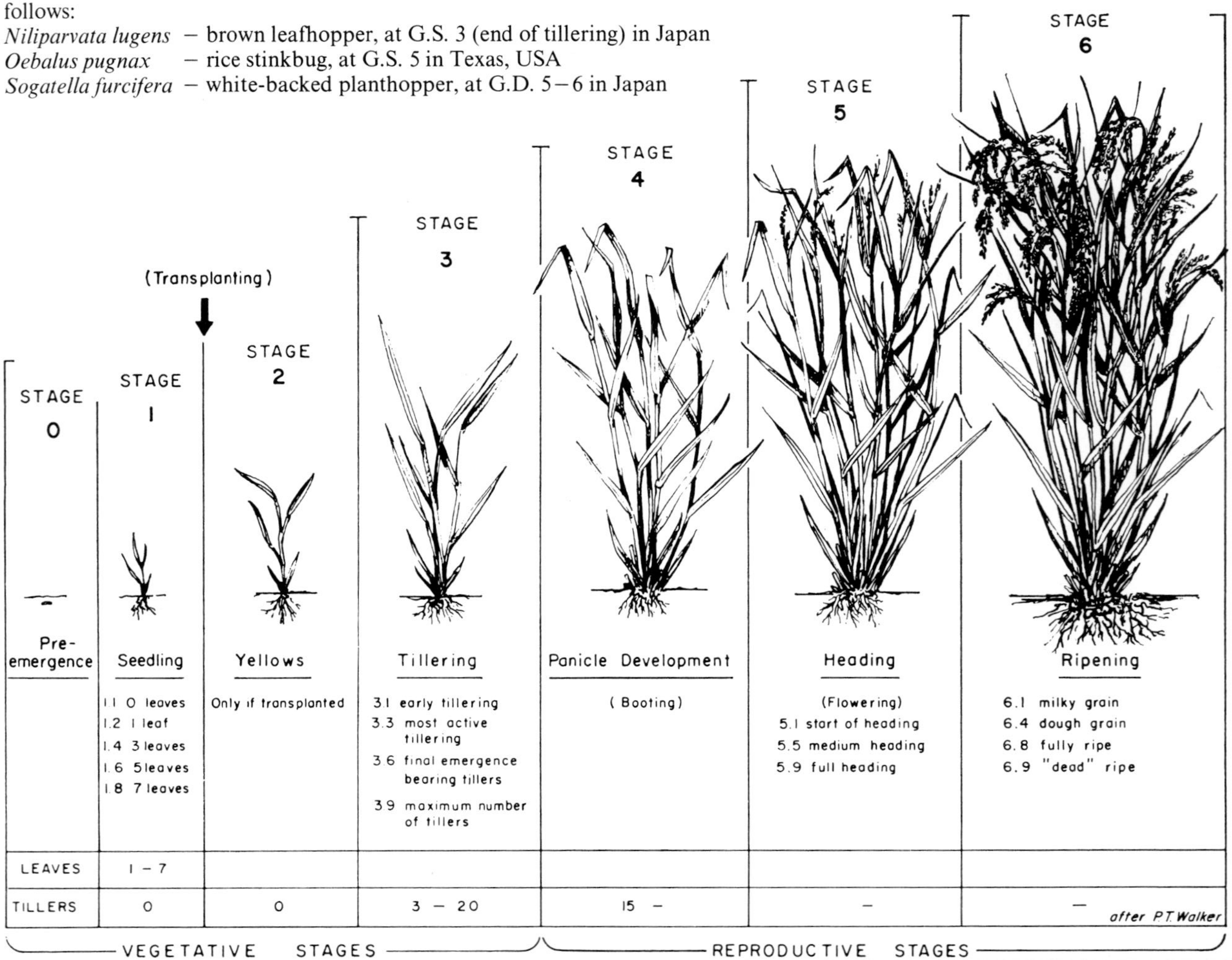

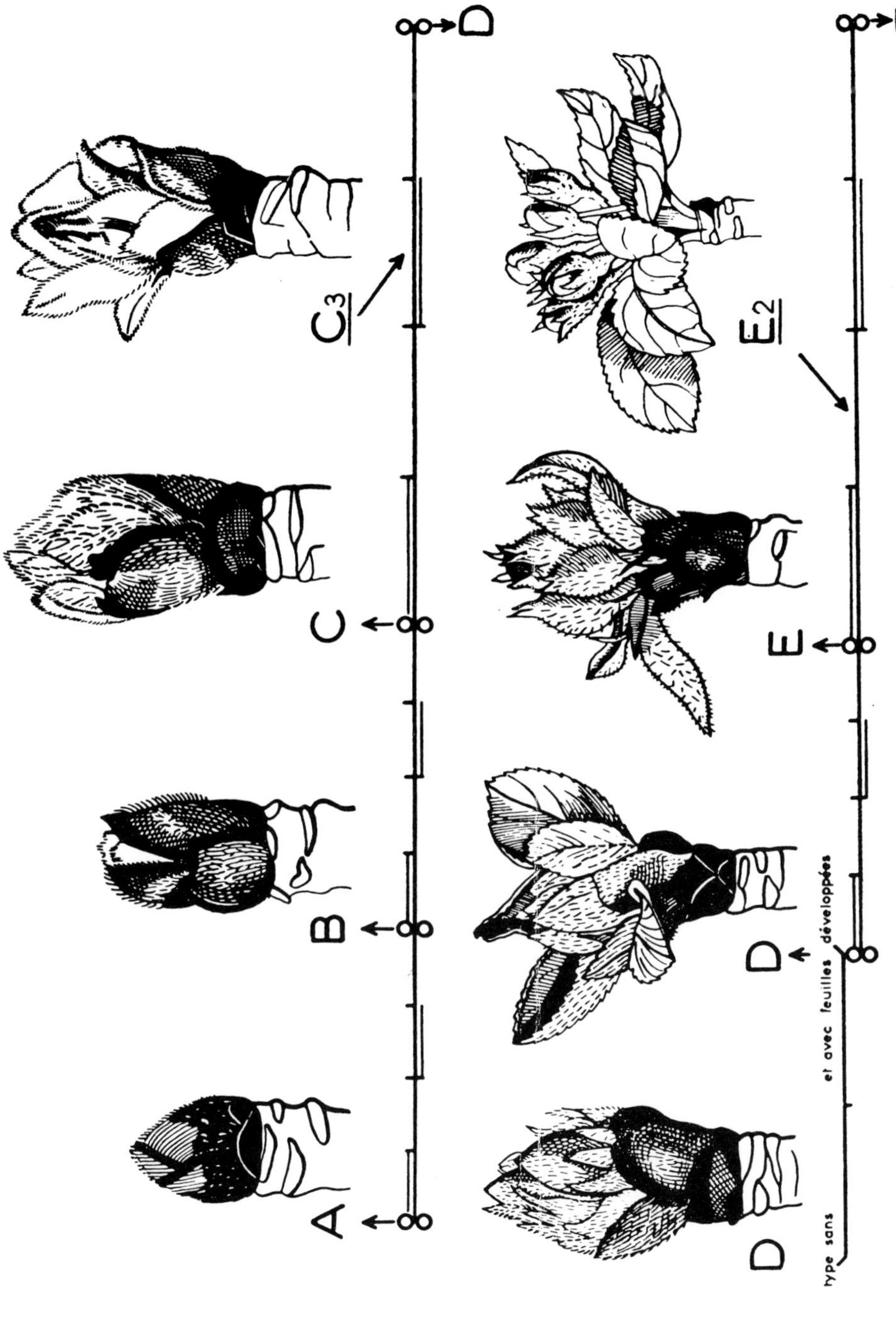

Fig. 4. Growth stages of apples (Anonymous 1952)

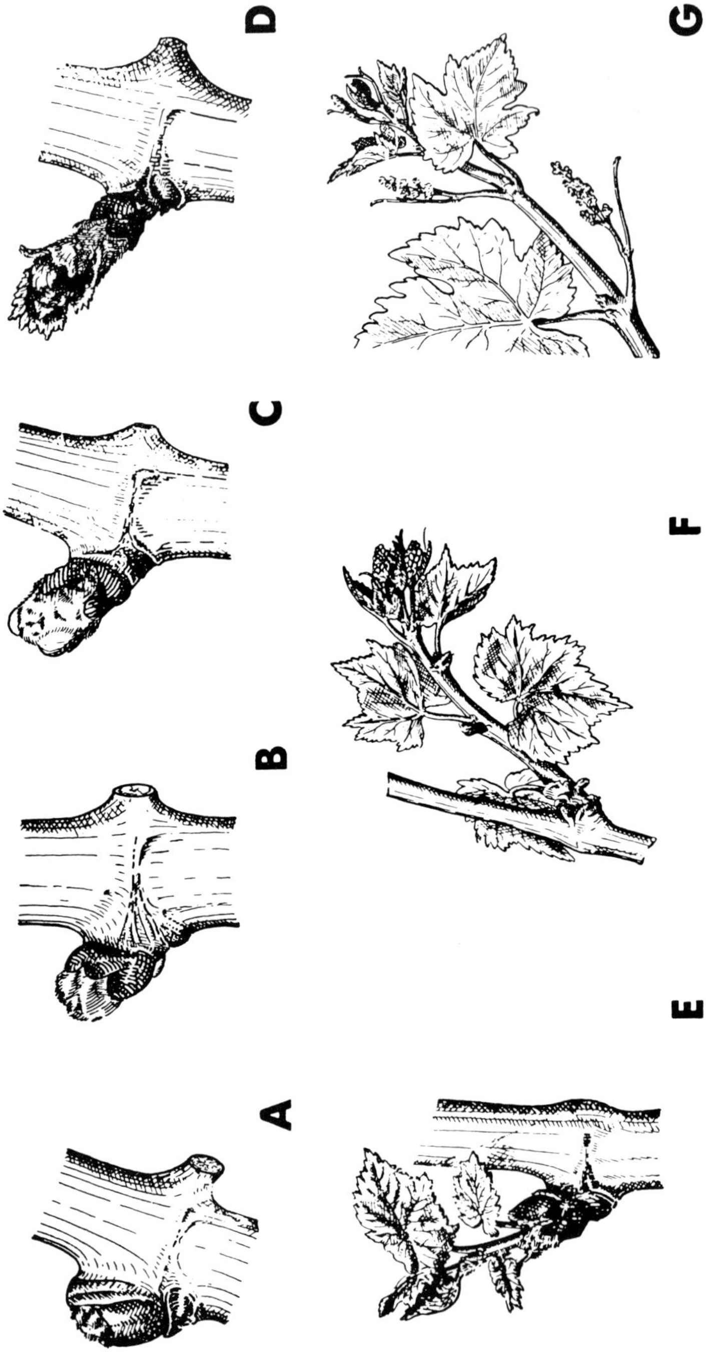

Fig. 5. Growth stages of grape-vine (Anonymous 1952)

Excellent classifications of the growth stages of deciduous fruit trees and vine, for use in relation to disease outbreaks, have been published in Italy, Switzerland and France. Those reproduced in Figs. 4 and 5 have appeared in a publication of the Service de la Protection des Végétaux, in France (Anonymous 1952). Two examples may be given: Apple scab (*Venturia inaequalis*) is not expected to develop before the sprouting has reached the stage designated as C_3 (see Fig. 4). It is only at or after this stage that scab attack is imminent if and when weather conditions favour the release of ascospores from overwintering perithecia.

As for the downy mildew of vine (*Plasmopara viticola*) the risk of primary infection commences as growth reaches stage F (Fig. 5), when shoots are about 10 cm long and have four to six expanded leaves, and the rudiments of grape bunches can be discerned. At this stage rainfall of about 10 mm combined with mean daily temperatures above 10 °C for 2–3 days will cause infection.

References

Anonymous (1952) Avertissements agricoles. Bull Tech Stat Avertissements Agric, 6 pp

Anonymous (1968) Plant disease development and control. Natl Acad Sci, Washington DC, Publ No 1596

Benedek P, Ebert W, Tanski VI, Dirlbek J (1981) Aufgaben der Prognose und Methoden der Entscheidungsfindung auf Betriebsebene. In: Überwachung und Prognose – Grundlagen eines gezielten Pflanzenschutzes. Akad Landwirtschaftswiss DDR, pp 220–231

Chiarappa L (ed) (1973) Crop loss assessment methods: FAO manual on the evaluation and prevention of losses by pests, diseases, and weeds. Commonwealth Agric Bur

Gutsche V, Kluge E (1983) Phyteb-Prognose, ein neues Verfahren zur Prognose des Krautfäuleauftretens (*Phytophthora infestans* Mont. de Bary). Nachrichtenbl Pflanzenschutz DDR 37:45–48

IOBC (Int Org Biol Control) (1980) In: Baggiolini M et al. (eds) Visuelle Kontrolle im Apfelanbau. IOBC/WPRS Heft 2

Jeger MJ, Butt DJ (1983) The effect of weather during perennation on epidemics of apple mildew and scab. EPPO Bull 13:79–85

Kable PF (1978) Practical management of crop disease. Dep Agric NSW

Mackenzie DR (1981) Scheduling fungicide applications for potato late blight with BLITECAST. Plant Dis 65:394–399

Makarova LA, Poljakov IJ, Grala B, Gutsche V, Pohle H (1981) Phenologische Prognose. In: Überwachung und Prognose – Grundlagen eines gezielten Pflanzenschutzes. Akad Landwirtschaftswiss DDR, pp 155–166

Rice RE, Flaherty DL, Jones RA (1982) Monitoring and modeling San Jose scale. Calif Agric 36:13–14

Rondomanski W (1969) Methode zur Prognose der Intensität der Primärinfektion durch den falschen Mehltau der Zwiebel in Polen. Arch Pflanzenschutz 5:89–96

Sevacherian V, Toscano NC, Van Steenwyk RA (1978) Heat accumulation method aids cotton pest management. Calif Agric 32:25

Turl LAD (1983) The effect of winter weather on the survival of aphid populations on weeds in Scotland. EPPO Bull 13:139–143

Ulrich J, Schroedter H (1966) Das Problem der Vorhersage der Kartoffelkrautfäule (*Phytophthora infestans*) und die Möglichkeit einer Negativprognose. Dtsch Pflanzenschutzdienst 18:33–40

Wedberg JL, Ruesink WG, Armbrust EJ, Bartell DP, Steffey KL (1980) Alfalfa weevil pest management program. Coop Ext Serv, Univ Illinois, Circ 1136

Wetzel T, Spaar D, Mueller HJ (1982) Stand und Entwicklung zum integrierten Pflanzenschutz im Getreidebau der Deutschen Demokratischen Republik. Wiss Beitr Martin-Luther Univ, Halle-Wittenberg RS, No 33, part 1, pp 12–39
Wilson LT, Barnett WW (1983) Degree days: an aid in crop and pest management. Calif Agric 37:4–7
Young HC Jr, Prescott JM, Saari BE (1978) Role of disease monitoring in preventing epidemics. Annu Rev Phytopathol 19:263–285
Zalom FG, Goodell PB, Wilson LT, Barnett WW, Bentley WJ (1983) Degree-days: the calculation and use of heat units in pest management. Div Agric Nat Resources, Univ Calif, Leaflet 21373

2.4.3 Monitoring by Trapping Potentially Harmful Organisms to Guide Pesticide Applications

J. PALTI

2.4.3.1 Factors Affecting the Use of Insect Traps

Trapping potentially harmful organisms serves to guide control operations if these organisms are either passively wind-borne or actively flying or crawling, or both. Fungal spores and very small insects and mites are wind-borne and can then be caught, but cannot be attracted. The mobile stages of numerous insects can be attracted by many things that appeal to their need for food, shelter, and mating, and play on their senses of sight or smell. Food baits often take the form of decaying organic matter, e.g., fishmeal to attract flies; sight traps use the attraction of light to moths, beetles and other photo-sensitive insects, and the predilection of aphids, flies, and others to specific shades of yellow, while traps acting by smell most commonly function as phagostimulants or sex attractants. The devices used to trap passively wind-borne organisms are chiefly suction traps and sticky surfaces. These are frequently combined with food, sight, and smell traps to prevent the insects attracted from escaping. The colour of pheromone traps can also be of importance: males of the lilac borer (*Podosesia syringae*) are more strongly attracted to black, brown, or red than to white pheromone traps (Timmons and Potter 1981).

The simplest use of traps is to indicate whether or not harmful organisms are present. In the case of pests with clearly separate generations, this may serve to define pest-free periods (or at least periods in which certain pests are not on the wing), and to indicate when field inspections are not required and when they should begin. No quantitative aspect is then involved, and there is no problem in the interpretation of catches. This applies to many insect pests, and also to the trapping of early-season spores, such as ascospores released from the overwintering bodies of the apple scab fungus (*Venturia inaequalis*).

However, where the *size* of the catch is meant to indicate damage or treatment thresholds, the question of interpretation arises and requires careful consideration of the environmental conditions under which the traps operate, their design and positioning, and of the flight and mating habits of the insects concerned. Some of the factors to be considered are discussed below.

1. *Environment.* Most weather factors have some effect on the size of insect catches by all types of traps, e.g., catches in all traps are affected by temperature and rain. Wind speed is of special importance where suction traps are used, and the catches made by light traps have to be considered in relation to the colour and intensity of neighbouring light sources and of the brightness of the moon (Nowinszky et al. 1979), as well as to the strength of geomagnetic fields (Kiss et al. 1981).
2. *Design of Traps.* The size and shape of the trap largely determines its efficiency, especially in the case of light traps, e.g., whether or not the traps have "wings" to guide insects into the trap's container. Of special importance may be the size and shape of the orifice.
3. *Positioning of Traps*

a) *In or above the crop?* For suction traps it has been found that the higher the trap is placed above the crop level, the more likely is it to catch small migratory pests flying in day-time, such as aphids, rather than certain flies (Lewis 1980). Colour and light traps are the more efficient, the better they are placed to be seen by photo-sensitive insects: yellow dishes placed at the level of the top of oilseed rape and set to face the direction of prevalent winds, have been found most efficient in indicating colonization flight of flea beetles, weevils, and pollen beetles affecting this crop in France (Lechapt 1980).

b) *At the edge or inside fields?* Placement of male-baited pheromone traps of boll weevils (*Anthonomus grandis*) around cotton fields has been reported to be much more effective than their placement in the field (Jacobson 1972). This is equally true for some moths, such as *Spodoptera littoralis* and *S. litura,* that affect cotton (Neumark and Teich 1980).

4. *Size of Insect Populations.* In the case of pheromone traps, the size of the catch often indicates the size of insect populations more reliably when these are small than when they have reached large numbers. The pheromone traps have to compete with the attraction of virgin males or females, and can do this more effectively as long as not too many virgins are in the vicinity of the trap. The fact that reliability of indications by catches in pheromone traps in apple orchards declines as insect populations increase has been stressed in relation to the codling moth (*Laspereysia pomonella*) in France (Touzeau 1980) and the red-banded leaf-roller in the United States (Jacobson 1972).

We are here concerned with the use of traps for monitoring harmful organisms, and for establishing, where possible, threshold values for catches. But we shall not go into the use of hormone traps as control measures in their own right, by removing most or all males or females from the pest population, or by confusing them and thus reducing chances of mating.

2.4.3.2 Trapping Methods for Major Groups of Flying or Air-Borne Pests

In the following sections the methods by which insects belonging to major groups of pests can be trapped are briefly reviewed, and one example is detailed for each group.

Aphids and Whiteflies. Aphids can be trapped by light traps, but this applies only to species that fly at dusk or during the night, and does not give very reliable indications. The traps commonly used for catching aphids are suction traps or yellow colour traps, with sticky surfaces or water to catch the insects. With a few exceptions, aphids that feed on dicotyledonous hosts are attracted by yellow, while those feeding on grasses are not (Taylor and Palmer 1972). Yellow traps have the advantage that they catch more aphids and whiteflies than insects of any other groups. Various shades of yellow paint, made up of different bases, are in use. Sticky yellow colour traps have also been used with success in Israel to catch the tobacco whitefly (*Bemisia tabaci*) (Melamed-Madjar et al. 1981).

In a comprehensive scheme for catching aphids by suction traps, 24 such traps are operated in the United Kingdom, 10 in France, 3 in the Netherlands, and 2 in Belgium (Lewis 1980, Tatchell 1982). The insects are sampled daily at a height of 12.2 m, and the catches are analyzed weekly at the Rothamsted Experiment Station. This has proved cheaper and quicker than crop sampling, and provides an earlier warning of the presence of aphids. The catches are most useful for regional monitoring of aphids and for the timing of field sampling, but cannot substitute for the latter. The suction traps are used to monitor about 30 species of aphids, the most important being the cereal aphids, such as *Rhopalosiphum padi*, and sugarbeet aphids, such as *Myzus persicae* and *Aphis fabae*. The scheme works as follows: results of the samplings are transferred to a microcomputer, where they are validated and used to produce two weekly publications, the *Aphid Bulletin* which records the numbers of each species caught during the week, and the *Aphid Commentary*, in which catches of nine species are compared with information previously collected, to form the basis of forecasts. This seems an excellent example of a service developed and run by a research institution, operated to aid farmers and advisory personnel. A combined trapping and radar network to monitor aphid movement is now being developed.

Scale Insects. The important citrus pest, the California Red Scale (*Aonidiella aurantii*), can readily be caught by synthetic hormone traps. The trap used in Israel has a triangular cross-section, with a changeable sticky sheet on the floor, and a rubber dispenser for 100 mg pheromone. The latter remains active for more than 4 months under the hot spring and summer conditions of the Jordan Valley, where one trap per hectare is used (Sternlicht et al. 1980). In California, after the first flights of this scale have been trapped in April – May, temperature records are used to indicate the rate at which generations develop (see Sect. 2.4.2.3, Table 3).

The San Jose scale (*Quadraspidiotus perniciosus*) has been routinely monitored in California by use of the Zoecon SJS "tent" trap, hung 1.8–2 m high in the north or east side of trees (Rice et al. 1982). In orchards larger than 12 ha, one trap per 8 ha is used, but never less than a total of three traps, because catches in individual traps are so variable. The synthetic hormone is replaced every 6 weeks, and traps should be inspected twice weekly.

Flies and Midges. Adult flies can be attracted in a great variety of ways. The attractiveness of protein hydrosylates to the Mediterranean Fruit Fly *Ceratitis capitata*, is widely used for monitoring and controlling this pest. Traps based on ammonium salts are in use to monitor *Rhagoletis* flies that attack cherries (Prokopy and Hauschild 1979) and walnuts (Riedl et al. 1981) in the United States. Light traps have been reported effective in catching the wheat bulb fly (*Delia coarctata*) in England (Bowden and Jones 1979), and pheromone traps are among those proposed to catch the olive fly (*Dacus oleae*).

However, most fly traps used in agriculture base on the attraction of colours and on trapping attracted flies by sticky materials. Dark red spheres have proved highly attractant to the cherry fly, *Rhagoletis pomonella* (Prokopy and Hauschild 1979). Sticky green spheres were found the best indicators of egg laying by *R. completa*, the walnut fly (Riedl et al. 1981).

Most frequently used are yellow colours, and the exact shade is highly important. Following the pioneering work published by Russ et al. (1973) in trapping the cherry fly in Switzerland, Hameiri (1973) introduced colour baiting in Israel to monitor, and in some cases control, the olive fly (*Dacus oleae*) (see Chap. 3.3); the shade of yellow most effective was the one designated as RAL 1004. This was painted on cardboard plates covered with a glue that was completely transparent, remained sticky for 2–3 months in the grove, and did not alter the color of the plate. The traps were positioned in groups of five on alternate trees in test units consisting of five trees each, in three contiguous rows of trees, with one such unit serving 2–3 ha of olives. Monitoring appearance of the olive fly in this way halved the number of spray applications required for its control.

A technique for pest monitoring by collecting not adults but prepupae has been worked out by Johnson et al. 1980 for control of *Liriomyza sativa* on tomatoes. Styrofoam trays placed on the ground under tomato foliage proved an effective and inexpensive way of trapping prepupae and indicating changes in population.

An egg trap to guide pesticide use for the control of the cabbage fly *Delia radicum* has been developed in Switzerland (Freuler and Fischer 1983).

Beetles and Weevils. Beetles and weevils can be trapped in a great variety of ways, by yellow colour traps, black light traps, ordinary light traps and, in their crawling stage, by pitfall traps. All these traps serve to catch both sexes.

Most widespread is the use of yellow colour traps. Yellow light attracts the cotton boll weevil (*Anthonomus grandis*) (Leggett and Cross 1978), and in France yellow dishes have been used extensively for trapping flea beetles (*Phyllotreta* spp.), weevils (*Ceutorrhynchus* spp.) and pollen beetles (*Meligethes* spp.)

in oil seed rape (Lechapt 1980). Flight of these pests can be detected over a wide area with a small number of traps. The risk of damage to the oil seed rape by each pest has been closely related to the crop's developmental stages, which have been divided into 18 strictly defined growth stages (from A = cotyledon stage, to G_5 = grain ripening); the crops are most susceptible to damage by pollen beetles in stages D_1 (flower buds still covered by terminal leaves) to E (elongation of flower peduncle). Thresholds have been worked out for the number of insects caught in one day, e.g., 10 *Ceutorrhynchus* weevils, or over several days in cumulative trapping, e.g., 30 such weevils.

Pitfall trapping at the edge of lucerne and clover fields has given good indications of seasonal movement of adults of *Sitona hispidulus* weevils in Kentucky, U.S.A. (Leibee et al. 1981).

Sticky plastic sheet traps, placed under the canopy of trees populated with the chafer *Rhizotrogus majalis*, have proved effective and cheap in capturing the beetle in Canada (Fiori 1983).

Moths and Butterflies. Photo-positive moths and butterflies can be trapped by light or pheromone traps. Ordinary and black light traps catch a wide range of insects, both harmless and injurious, of both sexes; they therefore provide an insight into the size as well as the species and sex ratio of the insect fauna in flight at any given time, which is useful for surveys. But their use has two important drawbacks: (a) a source of power is needed to make them work, and they require frequent attention, and (b) sorting of the large number of different insects trapped indiscriminately requires much time of trained personnel. Light traps are therefore used less and less for monitoring Lepidoptera, and in their place sex attractant traps are used wherever available.

Traps containing virgin females have been used for years to catch the males of various moths and butterflies, but they have the obvious disadvantage that rearing the females, maintaining the traps, and constantly adding new females, requires skilled work. Thus extensive progress in using sex attractants has come about only since considerable numbers of sex hormones have been synthesized successfully and have become commercially available at reasonable prices.

The widest use of such traps has probably been made in the monitoring and control of the codling moth (*Laspeyresia pomonella*) in apple orchards in Europe and North America. Methods of trapping this pest in France have been detailed by Audemard (1979). He defined the factors influencing the size of catches in the triangle reproduced in Fig. 1. This shows that the reliability with which codling moth control can be guided by the size of catches depends on numerous factors, of which we would like to single out the following two: (a) source of the moth population: if there is a large influx from outside the orchard, the size of catches can be erratic, and the number of males caught does not necessarily indicate the population of female moths; and (b) temperature: since temperatures above 15 °C are required for mating and oviposition, spring catches of the first generation may not be indicative of attack, while catches of the second generation, developed at higher temperatures, are more reliable indicators. In this and other work in France it has been found that one pheromone trap could service up to 10 ha of apple orchards.

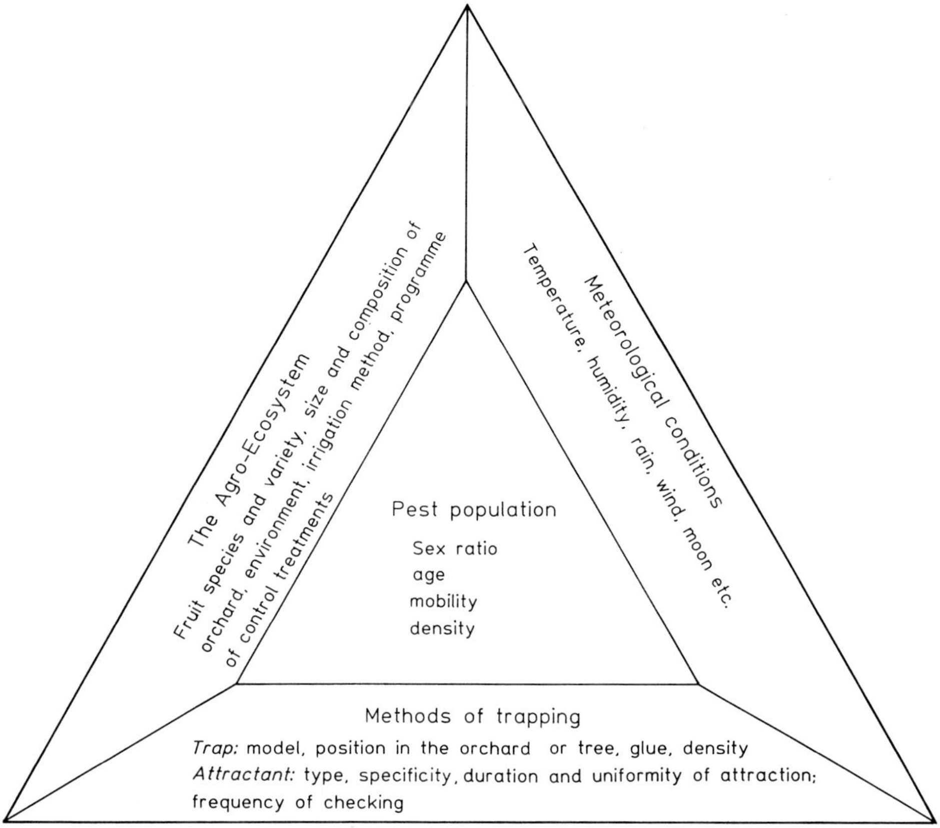

Fig. 1. The factors affecting the size of catches of the codling moth (*Laspeyresia pomonella*) in pheromone traps. (After Audemard 1979)

A few other aspects merit consideration to facilitate interpretation of the size of catches by pheromone traps (IOBC et al. 1980): (a) In some species of moths the males hatch earlier than the females; thus, in Southern Germany only males of the codling moth are commonly caught up to mid-June, and only monitoring of oviposition (or use of light traps) will then indicate when control operations may have to begin. (b) Cold evenings or strong winds depress catches; where such conditions prevail, trapping has to be supplemented by search for eggs. (c) Although pheromones are theoretically species-specific, males of other moths which are often harmless are commonly caught in pheromone traps, sometimes making up 20% of the total catch; only careful and discerning evaluation of the catch will then prevent erroneous interpretation.

Two technical points to increase catches of moths by pheromones are of interest. The fact that olfactory and visual stimuli could be combined and that the colour of the trap may be important in capturing moths by pheromones has been demonstrated by Timmons and Potter (1981) in Kentucky, U.S.A. As mentioned above, they caught significantly more male moths of the lilac borer

(*Podosesia syringae*) in black, brown, or red pheromone traps than in those of any other colour, with white traps giving poorest results.

Mechanical improvement of the trap may also be worthwhile. Coating the polyethylene funnel of a pheromone trap with kaolin or calcium caused increasing numbers of *Spodoptera littoralis* male moths to slide into the trap, raising the catch by up to 60% (Navon 1978).

2.4.3.3 Trapping Spores of Fungi

Spore trapping is not so far being used extensively by crop protection advisers or farmers. The equipment is not cheap and needs maintenance; but the principal reason why spore traps have not been widely adopted in plant protection practice is that interpreting the catches requires considerable expertise: spores of many mildews and many rusts look much alike, and the identity of the fungus producing them is not easily determined. This leaves a regrettable vacuum in the field of disease prediction: ascertainment of the presence and quantity of spores is an important component of prediction of diseases caused by airborne pathogens. In fact, in the presence of large amounts of their spores, some pathogens have been shown to be able to overcome weather or host resistance handicaps, i.e., the mass of spores compensates the pathogen for unfavourable parameters, such as short periods of leaf wetness (Aust et al. 1980).

The most important disease in the monitoring of which spore trapping plays a practical role is the scab of apples and pears. Zuck and MacHardy (1981) have reported on the development of an inexpensive volumetric spore trap to assist advanced growers in Maine (U.S.A.) in assessing the inoculum levels of apple scab (*Venturia inaequalis*) in each orchard. The trap is semi-automatic in operation, using a hemp string as the basis for a moisture-activated switch that can be coupled to an electronic timer to allow variable sampling periods. Power is supplied from a car or motorcycle battery, and spores are counted under the microscope.

Other pathogenic fungi, the control of which can be guided by trapping their spores include the downy mildews of hop (*Pseudoperonospora humuli*) and of tobacco (*Peronospora tabaci*) and the fungus causing canker of pome fruits (*Nectria galligena*) (see also Chap. 3.2).

2.4.3.4 Monitoring Viruses and Their Vectors

The subject of monitoring virus vector populations and the incidence of the viruses they carry has recently been reviewed by Raccah (1983). The procedure is complex, because to be reliable it involves two steps: assessing the size of the insect population that can carry the virus and is apt to land on the crop, and subsequently determining the proportion of these insects that actually do carry the virus.

To assess how many potential vectors alight on the crop, traps should be used that are just about as attractive to the insects as the crop itself, but not markedly more attractive, as is the case with some yellow traps. To this purpose, Irwin (1980) has designed an aphid trap consisting of a Perspex receptacle containing a horizontal tile coloured with the shade of green of soybean foliage, as determined accurately by spectrophotometry. The number and identity of aphids caught in this trap and that of aphids landing on the vegetation has been found to be closely similar.

The second step, determination of the infectivity of the vector population, is more difficult, but many crop protection advisers should be able to carry it out. This involves exposure in the field, for a few hours, of batches of some 10–20 potted plants. After their removal from the field, the plants are kept under observation for 1–2 weeks. The proportion of plants that then shows symptoms of infection, in conjunction with the number of potential vectors trapped, indicates the danger to the crop of imminent virus infection.

The above method, use of green tile traps and bait plants, has proved reliable where at least 5% of the vector population carries the virus, and where plants are developed sufficiently to cover most of the soil surface. It has been used successfully to monitor the occurrence of Soybean Mosaic Virus (SMV) on soybean in Illinois (USA) and of Cucumber Mosaic Virus (CMV) and Potato Virus Y (PVY) on pepper in Israel.

Monitoring virus diseases is of great importance. Very little can be done in the field to curb their spread from first foci of infection. Farmers therefore depend on early warning services to estimate their chances of obtaining a worthwhile crop from the field in question. The procedure involved in such monitoring is more complicated than mere trapping or counting of parasites. However, in view of the benefits that may be derived from this operation, crop protection advisers should be trained to carry it out wherever necessary.

References

Audemard H (1979) Le piégage du carpocapse (*Lespeyresia pomonella* L.) avec la pheromone sexuelle de synthèse Dodecadiene 8E, 10E O11. Defense Veg 33:29–43

Aust HJ, Bashi E, Rotem J (1980) Flexibility of plant pathogens in exploiting ecological and biotic conditions in the development of epidemics. In: Palti J, Kranz J (eds) Comparative epidemiology. PUDOC, Wageningen, pp 46–56

Bowden J, Jones MG (1979) Monitoring wheat bulb fly, *Delia coarctata* Fallen (Diptera: Anthomyiidae) with light traps. Bull Entomol Res 69:129–139

Fiori BJ (1983) Sticky plastic sheet traps as survey tools for the European Chafer (Coleoptera: Scarabaeidae). Can Entomol 115:1429–1431

Freuler J, Fischer S (1983) Le piège à œufs, nouveau moyen de prévision pour la mouche du chou, *Delia radicum (brassicae)* L. Rev Suisse Vitic Arboric Hortic 15:107–110

Hameiri Y (1973) Colour traps for the olive fly. Isr Minist, Ext Serv Agric (Hebrew, with English summary)

IOBC (International Organization for Biological Control) (1980) Steiner H et al. (eds) Die Klopfmethode, mit einem Anhang über Licht- und Pheromonfallen. IOBC/WPRS Heft 4

Irwin ME (1980) Sampling aphids in soybean fields. In: Kogan M, Herzog DC (eds) Sampling methods in soybean entomology. Springer, New York Heidelberg Berlin, pp 239–259

Jacobson M (1972) Insect sex pheromones. Academic Press, London New York

Johnson MW, Oatman ER, Wyman JA, Steenwyk RA Van (1980) A technique for monitoring *Liriomyza sativa* in fresh market tomatoes. J Econ Entomol 73:552–555

Kiss M, Ekk I, Toth G, Szabo S, Nowinszky L (1981) Common effect of geomagnetism and change of moon phases on light trap catches of fall webworm (*Hyphantria cunca* Drury). Z Angew Entomol 91:403–411

Lechapt G (1980) Méthodes de determination des risques concernant les insectes nuisibles au colza. EPPO Bull 10:119–128

Leggett JF, Cross WH (1978) Boll weevils: The relative importance of color and pheromone in orientation and attraction to traps. Environ Entomol 7:4–6

Leibee GL, Pass BC, Yeargan KV (1981) Seasonal abundance and activity of *Sitona hispidulus* adults in Kentucky. Environ Entomol 10:27–30

Lewis T (1980) Britain's pest monitoring network for aphids and moths. EPPO Bull 10:39–46

Melamed-Madjar V (1982) A method for monitoring *Bemisia tabaci* and timing spray applications against the pest in cotton fields in Israel. Phytoparasitica 10:85–91

Navon A (1978) A modified dry pheromone trap for *Spodoptera littoralis* male moths. Phytoparasitica 6:21–24

Neumark S, Teich I (1980) A dry and open pheromone trap activated by 4% diazinon for trapping male moths of *Spodoptera littoralis* Boisd. J Environ Sci Health A 15:313–321

Nowinszky L, Szabo S, Toth G, Ekk I, Kiss H (1979) The effect of the moonphases and of the intensity of polarized moonlight on the light-trap catches. Z angew Entomol 88:337–353

Prokopy RJ, Hauschild K (1979) Comparative effectiveness of sticky red spheres and Pherocon AM standard traps for monitoring apple maggot flies in commercial orchards. Environ Entomol 8:696–700

Raccah B (1983) Monitoring insect vector populations and the detection of viruses in vectors. In: Plumb RT, Thresh JM (eds) Plant Virus Epidemiology. Blackwell Sci Publ, Oxford

Rice RE, Flaherty DL, Jones RA (1982) Monitoring and modeling San Jose scale. Calif Agric 36:13–14

Riedl H, Hislop RG, Barnett WW, Coates WW, Fitch LB, Loos JL, Olson WH, Profits JC, Schreader WR (1981) New monitoring methods for the walnut husk fly. Calif Agric 35:21–22

Russ K, Boller EF, Vallo V, Haisch A, Sezer S (1973) Development and application of visual traps for monitoring and control of populations of *Rhagoletis cerasi* L. Entomophaga 18:103–116

Sternlicht M, Bar-Sakai Y. Seibush J (1980) (Use of the sex pheromone of the California Red Scale). Alon Hanotea, July 1980 (Hebrew), pp 661–684

Tatchell GM (1982) Aphid monitoring and forecasting as an aid in decision making. In: Austin RB (ed) Decision making in the practice of crop protection. British Crop Prot. Council Monograph 25, pp 99–112

Taylor LR, Palmer JMP (1972) Aerial sampling. In: Emden HF van (ed) Aphid Technology. Academic, London New York, pp 189–234

Timmons GM, Potter DA (1981) Influence of pheromone trap colour on capture of Lilac Borer males. Environ Entomol 10:756–759

Touzeau J (1980) Use of pheromone traps in agricultural warning systems. EPPO Bull 10:97–104

Zuck MG, MacHardy WE (1981) Recent experience in timing sprays for control of apple scab: equipment and test results. Plant Dis 65:995–998

2.4.4 Monitoring by Counting or Appraising Pest Populations or Crop Tissue Affected

J. PALTI

When all methods available for trapping pests or predicting their appearance have been employed, by far the most important means of in-growth monitoring is the determination whether or not a disease or pest has made its appearance, and how much of the crop has been affected. This is mostly done in one of two ways: by counting the numbers of insects or mites (in their various stages) present in the crop or the numbers of plants affected, or by visual estimation or appraisal of the proportion of crop plants or crop tissue affected by pest or disease. Such counts or appraisals have manifold uses, in regional surveys and individual fields, and in experimentation and advisory work, and the methods used vary according to the degree of accuracy required and the time at the disposal of the monitor. We shall here approach the subject principally from the aspect of monitoring to give farmers prompt advice as regards early detection of attack and attainment of properly defined treatment thresholds.

Monitoring techniques have to be simple and time-saving. They therefore have to concentrate (a) on major pests or diseases, and (b) on the tissues or organs most susceptible to attack, e.g., buds, young foliage, or root collars. Treatment thresholds must then refer to pest populations or tissue affected in these organs, especially when these are to be marketed, such as fruits, or have a particularly marked influence on yield such as the flag leaf of cereals (see Sect. 2.4.6.2)

2.4.4.1 Counting Pests or Affected Plants

Wherever counting is feasible, it is obviously preferable to appraising populations of pests or of affected plants, since appraisal has an element of subjective decision-making. Counting serves chiefly for the determination of pest populations on the aboveground organs of plants. But it is also used to indicate the percentage of plants showing symptoms on the whole or most of the plant, such as symptoms of virus infection or of pests or pathogens causing wilting, e.g., wireworms or fungal or bacterial wilting agents. Counts can also be made of soil-borne nematodes or fungal bodies (sclerotia), but this requires special equipment and expertise.

We have chosen two examples to illustrate the use of counting various stages of insects and mites in the monitoring of pests. Populations of beneficial insects can also be assessed in this way.

Monitoring Cotton Pests in Israel. Pests are a major constraint in growing irrigated cotton in Israel, and their control is expensive. To reduce cost, and to time sprays correctly, fields are monitored weekly. They are divided into "monitoring plots" of 5–15 ha, and insects are counted on one 1–2-m-long row sections per hectare, never less than five such sections. In counting larvae of *Heliothis armigera,* three classes are distinguished and evaluated differently: larvae of less than 0.4 mm are disregarded, while larger larvae are weighted by being assigned "larval units": larvae measuring 0.4–15 mm are given 1 unit each, larvae longer than 15 mm are given 3 units. The treatment threshold for this pest is 4 units per section monitored. Additional data for cotton pest monitoring are presented in Table 1, and new developments in cotton monitoring are discussed in Section 2.4.6.2.

Apple Pests in Central Europe. A variety of counting methods have been developed in Central Europe for monitoring apple pests and diseases, in the frame-

Table 1. Criteria and thresholds in monitoring certain pests in irrigated cotton in Israel (Melamed and Shoham 1977)

Pest	Plant part sampled	Size of sample	Threshold
White fly (*Bemisia tabaci*)	5–6th leaf from the top	50 leaves	8 nymphs per leaf
Leaf hopper (*Empoasca lybica*)	Full-grown leaves sampled at random	30 leaves	6 nymphs per leaf
Bollworm (*Heliothis armigera*)	Whole plant	2 m section of row	4 larval units (see Sect. 2.4.6.2)
Spiny bollworm (*Earias insulana*)	Whole plant	2 m section of row	a) Until 2 bolls per plant have formed: 6 larvae b) At subsequent stages of growth: 2 larvae
Egyptian cotton leafworm (*Spodoptora littoralis*)	Whole plant	100 m row	8 egg-masses, with an average of 15 or more larvae surviving in each egg-mass

Table 2. Determination of populations of some apple pests by the beating-tray method and treatment thresholds at various seasons (numbers collected from 100 branches). (Adapted from IOBC 1980a, b)

Pest	Seasonal thresholds		
	Before blossoming	After blossoming	In summer (after July)
Operophthera brumata winter moth	8 – 12 small larvae	12–15 larvae	
Orthosia incerta clouded drap moth	2 – 3 larvae	3–5 larvae	4–6 larvae
Rhopalosiphum insertum apple-grass aphid	3000 aphids	3000 aphids	
Aphis pomi green apple aphid	20 – 50 aphids	40–50 aphids	50–80 aphids
Psylla mali apple sucker	150 – 150 larvae	Too late for control	

work of instituting integrated pest control. These include in-situ monitoring on buds, foliage, and fruits, and counting pests knocked off branches in a precisely defined operation (beating-tray method, Klopfmethode) which has been described by Steiner and colleagues (IOBC 1980b) as follows: A rectangular funnel measuring 40 by 62 cm at the top, with sides of any smooth fabric and a collecting bottle at the bottom, is held below the branch to be tested. The

Fig. 1. Checking pests and beneficial insects in an apple orchard by the beating-tray method. (By courtesy of Dr. K. Hanuss, Landespflanzenschutzamt Rheinland-Pfalz, Mainz, Fed. Rep. of Germany)

branch is rapped smartly three times with a bamboo stick 70–80 cm long, bearing a foam rubber head (see Fig. 1). 100 branches selected at random are tested in each plot, preferably in the morning. The pests collected and killed in the bottle are then counted and the significance of their numbers assessed according to the thresholds determined for each season, before and after blossoming, and later in summer. Table 2 presents examples of a number of apple pests and the seasonal thresholds determined for them in Germany and Switzerland (IOBC 1980a).

2.4.4.2 Visual Appraisal of Symptoms

Visual appraisal of symptoms has to take into account both the incidence and the severity of whatever ails the crop. It is used for assessing diseases and those pests the symptoms of which are easier to determine than their numbers, e.g., pests causing leaf curling, bud mites, some stem borers, etc.

Appraisal is convenient where (a) symptoms are clearly defined, and (b) a single parasite is dominant. It is much more difficult where symptoms are not well defined, where several causes (parasitic or not) are potentially involved, and where tissue dries up or organs shed. In the latter case it may be necessary to assess separately the proportion of still green or yellowish tissue, on which symptoms can be discerned, and the proportion of leaf, flower or fruit that has dried up or been shed (see below).

Every appraisal that goes beyond the mere indication of presence or absence of diseases or pests and aspires to a reasonable degree of accuracy has to include indications of incidence as well as of severity of crop affliction. Incidence, also termed prevalence, or frequency, refers to the *number* of plant units affected, i.e., the number of leaves, fruits or whole plants affected, expressed as percentage of the total number of units appraised. "Severity" refers to the *area* of plant tissue affected, either in particular organs or whole plants, expressed as percentage of the total area of plant tissue appraised. Both concepts can also be used to indicate percentage of healthy units or areas of tissue, and where the attack is severe or due to a variety of causes, this is sometimes preferable. The problems of sampling to obtain reliable data on the incidence of pests and diseases are discussed in Chapter 2.4.1.

Assessment of Incidence. As mentioned above, where whole plants show symptoms of ailing, incidence can be counted and does not have to be appraised. Appraisal is restricted to those diseases or pests that affect individual organs, relies on scales of gradings that often include severity values, and its accuracy is much dependent on the sampling method adopted.

Assessment of Severity. The severity of crop affliction is considerably more difficult to assess than its incidence. Horsfall and Barratt (1945) were the first to point out that severity grades distinguished by the human eye are approximately equal divisions on a log scale and follow the law that states that visual acuity (discernment) depends on the logarithm of the intensity of the stimulus. They further note that, in making a percentage disease assessment, the eye actually assesses the diseased area up to 50%, and the healthy area above 50% of affected tissue.

James (1983) has succinctly described methods for assessing crop disease severity, and the following paragraphs are largely based on his paper. These methods can be conveniently separated into two types. The first is a descriptive key which describes plants with different levels of disease and assigns a category, grade or percentage infection to each description. It is generally agreed that little or no accuracy is to be gained by using more than seven to eight grades, and fewer grades are mostly sufficient for appraisal in the course of routine monitoring. The best known key of this type is that published by the British Mycological Society for assessing late blight of potatoes, which is reproduced in Table 3.

The second type of assessment key utilizes standard area diagrams, which typify the extent to which disease has developed either on small, simple units (leaves, fruits) or on large, composite units, such as branches or whole plants.

Small, Simple Units. Diseases affecting the leaf, stem, or fruit may be assessed by using the whole or portions of these organs as units of appraisal. If the dis-

Table 3. Assessment key for late blight of potatoes (*Phytophthora infestans*) (Anonymous 1947)

Blight (%)	Nature of infection
0.0	No disease observed
0.1	A few scattered plants blighted; no more than 1 or 2 spots in 12-yard radius
1.0	Up to 10 spots per plant; or general slight infection
5.0	About 50 spots per plant; up to 1 in 10 leaflets infected
25	Nearly every leaflet infected, but plants retain normal form; plants may smell of blight; field looks green although every plant is affected
50	Every plant affected and about 50% of leaf area destroyed; field appears green, flecked with brown
75	About 75% of leaf area destroyed; field appears neither predominantly brown nor green
95	Only a few leaves on plants, but stems green
100	All leaves dead, stems dead or dying

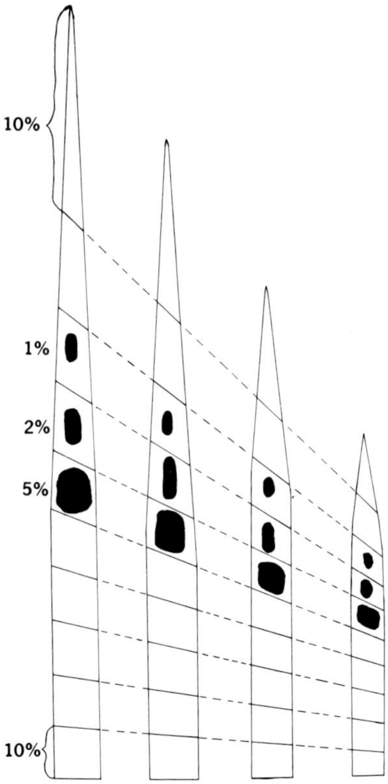

Fig. 2. *Rhynchosporium* assessment key. Each division represents 10% of the area of each leaf. *Black areas* represent 1%, 2%, and 5% of each leaf (James et al. 1968)

ease affects either surface of the leaf independently of the other, the elementary unit may be each surface of each leaf. Each grade of appraisal denotes a definite maximum proportion of affected plant tissue. James et al. (1968) have published the key depicted in Fig. 2 for assessing *Rhynchosporium secalis* leaf blotch on the two top leaves of barley. Instructions for using the key are given in the caption. To obtain an accurate estimate with the use of small units, large numbers of such units must usually be sampled, and the actual appraisal can then often be done in a convenient room or laboratory. But the procedure is laborious, and appraisal by small units is therefore not very often employed in routine monitoring work.

Large, Composite Units. Branches, vines, or tillers, and in many cases whole plants, parts of plots, or entire plots, represent large units of appraisal with components such as leaves or fruits, heterogenous in many respects and differing in age, position on the plant (inside or outside of a tree), and exposure to weather factors (dew). In such composite units there is usually an interdependence between the number of component organs affected (incidence) and the proportion of tissue affected in those organs (severity). If, for example, the majority of leaves of a plant is slightly affected, then we may expect a more or less definite, but smaller, number of leaves to be either free of symptoms or to be moderately infected, while few, if any, leaves will be severely affected. The pattern and range of infection degrees found in a single composite unit depends not only on the above-mentioned age and weather factors, but also on the rapidity with which the pathogen develops symptoms: if this is rapid (potato late blight) many components will show similar infection degrees, but this is not the case with parasites that take longer to develop significant populations.

Chester (1950) has described a key for whole-plant assessment of the leaf blight of maize (*Drechslera turcica*) (see Fig. 3). This is a disease which develops rapidly, and therefore a plant with heavy infection of its leaves (grade 4) will have few, if any, leaves free from symptoms. The scale of grades used here takes account of the number of lesions (very slight to abundant) as well as their position on the affected plant (lower to upper leaves).

Estimating or appraising grades of severity invariably involves a certain amount of bias. Reproducibility of a monitor's ability to distinguish between different grades can be easily examined. He should make two or more appraisals of the same plot on the same day; if these deviate markedly from each other, the plot should be re-examined to find where the difficulty lies (Berger 1980).

The frequency of appraisals depends on the speed of disease or pest development. For diseases, Berger (1980) recommends as a general rule intervals corresponding to one half of the incubation period. He holds that most leaf-spotting diseases should be appraised once to twice a week, beginning early enough so that several such appraisals can be made before an incidence of 1% is reached.

Special problems arise where part of the foliage area has dried up or leaves have been shed. If the parasite produces symptoms discernible on dried or shed leaves, such as holes or leaf spots bearing fungal structures (pycnidia or similar bodies), such leaves can be sampled and the proportion of affected leaves can then be distinguished from those that have dried up due to unrelated reasons, foremost among them senescence and strong winds. But even where affected leaves can thus be singled out, this is too time-consuming to be practiced in monitoring for other than experimental purposes. Instead, all that can be done

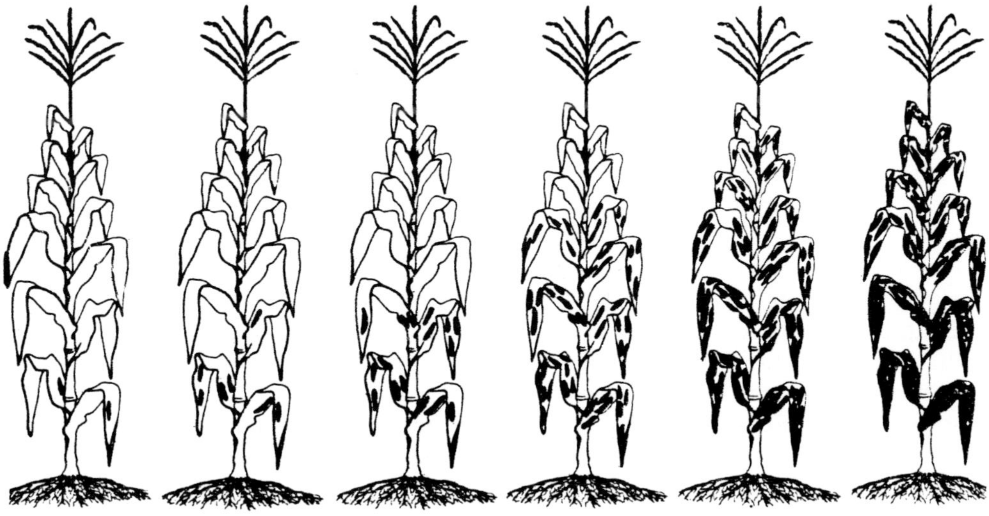

0.5	1	2	3	4	5
Very slight infection; one or two restricted lesions on lower leaves	Slight infection; a few scattered lesions on lower leaves	Light infection; moderate number of lesions on lower leaves	Moderate infection; abundant lesions on lower leaves, few on middle leaves	Heavy infection; lesions abundant on lower and middle leaves extending to upper leaves	Very heavy infection; lesions abundant on all leaves; plants may be prematurely killed

Fig. 3. Diagrammatic scale for appraising intensity of *Drechslera turcica* leaf blight on maize (Chester 1950). The leaf blight of maize is a disease that develops rapidly; this finds expression in the fact that a plant with heavy infection of its lower leaves (grades 3 and 4) will have few, if any, leaves free from symptoms. The scale of grades used takes account of the number of lesions (very slight to abundant) as well as their position on the affected plant (lower to upper leaves).

is an overall estimate of leaf desiccation or defoliation on the whole branch or tree. Unless the parasite to be appraised is so dominant that additional causes are of negligible importance, such estimates cannot aspire to great accuracy. But in many cases, such as the rust of stone-fruit trees (*Tranzschelia discolor*) or the powdery mildew of pepper (*Leveillula taurica*), defoliation is so important a symptom that any appraisal which disregards it will be seriously deficient (Palti 1949).

Monitoring with the use of whole plots as units of appraisal can be very useful for large scale surveys. But it is not, as a rule, specific enough for the type of monitoring demanded of the crop protection adviser whose aim is to obtain appraisals on which control prescriptions can be based. It is certainly not suitable for one of the primary aims, the detection of initial attack by pests or diseases.

Evaluation of Symptom Appraisal and Establishment of Thresholds. Since appraisal has potentially a larger margin of error than counting, the treatment thresholds derived from appraisal are of necessity less precise. In the case of parasites which predominantly or exclusively affect the yield directly, such as many fruit flies or rots, thresholds are derived from the loss in weight and grade expected when a given percentage of the yield organs is affected, and this is much the same as in the establishment of thresholds by counting.

But establishment of thresholds for pests or pathogens causing indirect damage, on the basis of the appraisal of affected tissue, is more complicated. The yield reduction caused mainly by pests or diseases restricted to the leaves of fruit or root crops is in many cases closely related to the power of recuperation of the plant and to other physiological and weather factors, as pointed out in Chapter 2.3. The conversion of appraisal data into treatment thresholds can then only be done on the basis of extensive local experimentation on a fairly uniform crop; such thresholds will apply only to a given variety, density, and sowing date, and not to crops grown in another region under different cultural conditions. This explains why treatment thresholds based on appraisal have to be evolved in conjunction with research institutions, and why use of such thresholds demands caution and expertise.

References

Anonymous (1947) The measurement of potato blight. Trans Br Mycol Soc 31:140–141

Berger RD (1980) Measuring disease intensity. In: Assessment of losses which constrain production and crop improvement in agriculture and forestry. Misc Publ 7, Agric Exp Stu Univ Minnesota, pp 28–31

Chester K, Starr (1950) Plant disease losses: their appraisal and interpretation. Plant Dis Rep Suppl 193:191–362

Horsfall JG, Barratt RW (1945) An improved grading system for measuring plant diseases. Phytopathology 35:655

IOBC (Int Org Biol Control) (1980a) In: Baggiolini M et al. (eds) Visuelle Kontrolle im Apfelanbau. IOBC/WPRS Heft 2, 96 pp

IOBC (Int Org Biol Control) (1980b) In: Steiner H et al. (eds) Die Klopfmethode. IOBC/WPRS Heft 4, 142 pp

James WC (1983) Crop loss assessment. In: Johnston A, Booth C (eds) Plant pathologist's pocketbook. Commonwealth Mycol Inst, Kew, Surrey

James WC, Jenkins JFE, Jemmett JL (1968) The relationship between leaf blotch caused by *Rhynchosporium secalis* and losses in grain yield of spring barley. Ann Appl Biol 62:273–288

Melamed Y, Shoham Ch (1977) Insect scouting in cotton fields in Israel. Minist Agric Ext Serv, 17 pp

Palti J (1949) Methods of assessing the incidence of olive leaf spot. A contribution to the technique of plant disease estimation. Palestine J Bot Rehovot Ser 7:156–166

2.4.5 Remote Sensing by Aerial Infrared Colour Photography as an Aid in Monitoring Crops for Pests and Diseases

C. H. BLAZQUEZ, R. HOCHBERG, and G. EDELBAUM

2.4.5.1 Principles of Remote Sensing

Remote sensing uses the properties of the electromagnetic spectrum to survey and diagnose various phenomena in the environment. It is based on the principle that each body reflects or absorbs radiant energy as electromagnetic waves with specific properties. How these waves are reflected or absorbed depends on the properties and shape of each body, the type and intensity of energy it receives, and on various environmental conditions. The spectrum of electromagnetic waves has a wide range, and only some of its bands serve for gathering the information needed. Figure 1 indicates the visible and the infrared parts of the spectrum that can be measured with the technology presently available.

A distinction must be made between passive sensing, which employs the sun as source of energy, as in photography, and active sensing which transmits energy, as in the use of radar. In agriculture, passive sensing is usually employed, with wavelengths of 0.3 – 15 μm. In this range we can distinguish between reflected and emitted energy (Blazquez & Horn 1980).

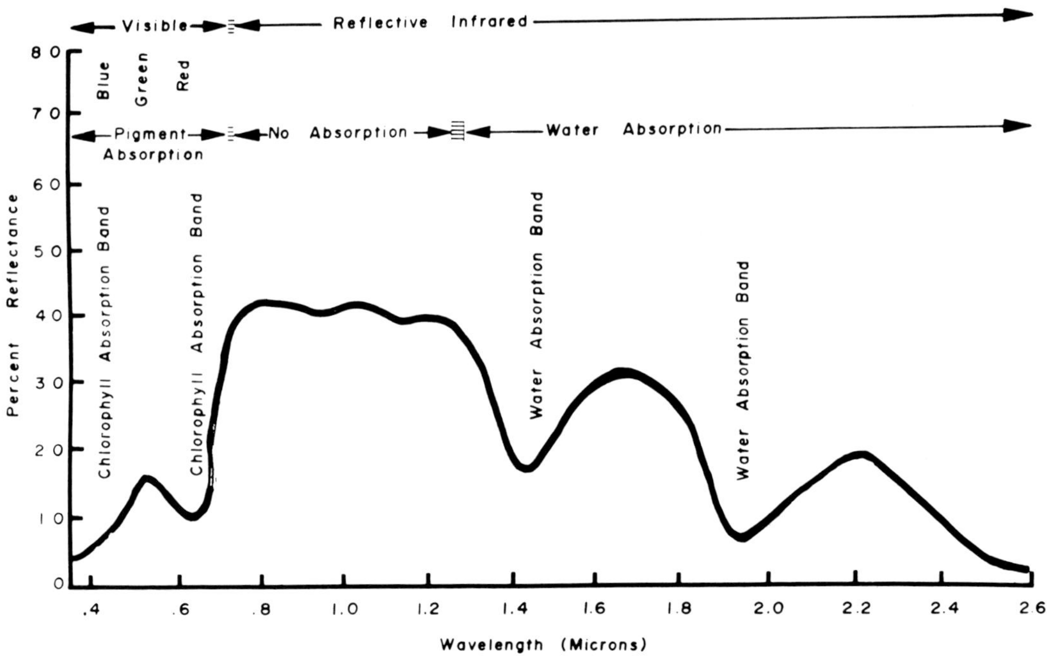

Fig. 1. Reflectance curve of a healthy, adult green plant with indication of chlorophyll and water absorption bands in the visible and infrared parts of the electromagnetic spectrum

Understanding the interaction between waves of various lengths and plants is essential for correct quantitative and qualitative interpretation of the data obtained. This interaction has therefore been studied widely on numerous crops, e.g., by Gates (1970) and by Knipling (1970). These studies have shown that the part of the electromagnetic spectrum which can be photographed is so far the most useful in crop protection work, and that infrared colour photography has specific advantages. While the sensitivity of ordinary colour photography parallels that of the human eye, infrared photography gives visual expression to a narrow range (0.7 – 0.9 μm) beyond the range of the eye, and thus makes it an effective tool for diagnosis. Sun radiation reflected by foliage is what infrared colour photography records.

2.4.5.2 Factors Affecting Reflectance from Crops

The factors affecting reflectance from plant tissue may be divided into three groups:

a) Physiological factors: pigments, water content and nutrient level.
b) Morphological factors: leaf structure, light refraction index within the mesophyll, leaf surface characteristics (cuticle, hairiness).
c) Spatial factors: leaf area index (see Chap. 2.5), crop density, growth habit.

Since these factors determine reflectance from all plant tissue, the reflectance curves for all crops run similar courses, with differences between them caused by the intensity of reflectance (Fig. 2). Pigments and leaf morphology much affect relectance, but considerable (and sometimes decisive) importance also attaches to spatial factors, particularly to leaf area index and crop density.

Pigments. The pigments supporting photosynthesis are the principal reflectors of radiation in the visible range. Mere changes in pigmentation, e.g., chlorosis, can therefore be detected by ordinary colour photography, and this has been demonstrated by surveys of avocado and other plantations suffering from mineral deficiencies. Infrared photography is, however, superior in this respect, and has the added advantage that it also records changes due to morphological and spatial factors.

Leaf Structure. The structure of the leaf as a whole, and in particular of the mesophyll, plays a central role in determining the type and intensity of radiation reflected from plants in the near infrared range. In most plants the mesophyll, which is the basic photosynthetic tissue, occupies the greater part of the leaf volume. Mesophyll tissues contain air spaces of various size, which aid in gas exchange. Light in wavelengths of 0.75 – 1.35 μm is reflected according to the tissue's relative content of air, water, and cell walls. An adult, healthy leaf absorbs hardly any energy in this range, and radiation either passes through such a leaf or is reflected by it. However, when the mesophyll is being destroyed by a harmful agency, reflectance changes, and this shows up in the colour obtained on infrared colour film. Figure 3 shows reflectance curves from affected citrus foliage.

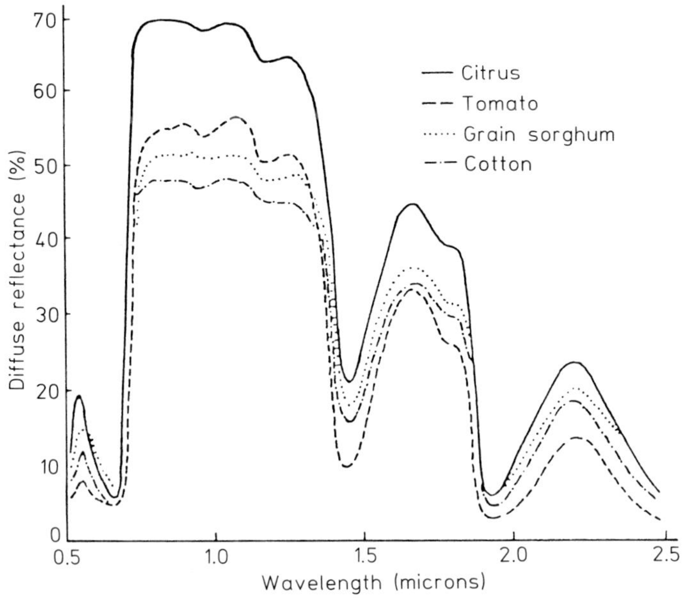

Fig. 2. Typical reflectance curves from leaves of various plants, as determined by spectroreflectometer in the laboratory

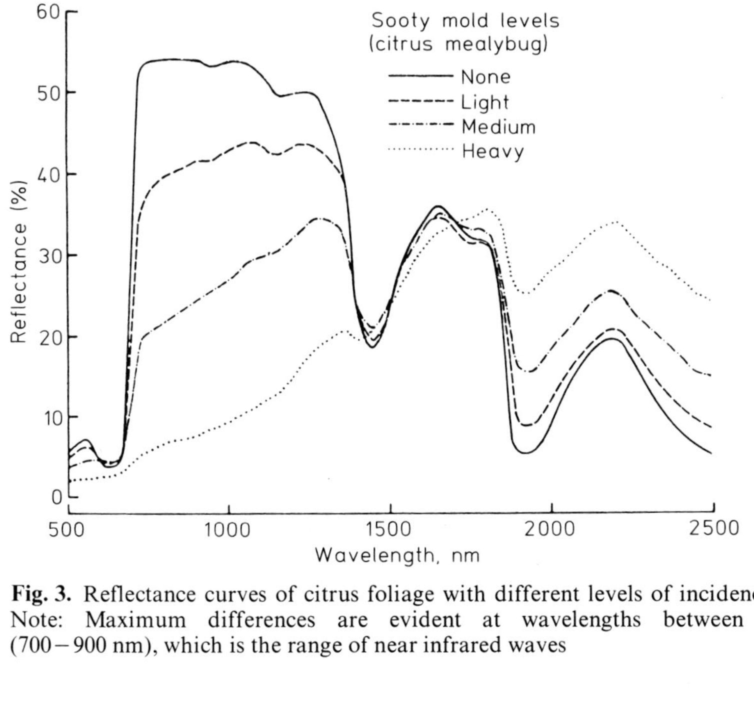

Fig. 3. Reflectance curves of citrus foliage with different levels of incidence of sooty mould. Note: Maximum differences are evident at wavelengths between 0.7 and 0.9 μm (700 – 900 nm), which is the range of near infrared waves

Spatial Factors. The reflectance of radiation from crops is greatly influenced by the "depth" of its foliage, i.e., the number of leaf layers superimposed on each other, which is an indicator of the crop's biomass. When crops are affected, by disorders, pests or diseases, they lose in depth, and this is readily recorded by infrared colour photography.

2.4.5.3 Criteria for Interpreting the Photographs

From the description of factors affecting reflectance, it is evident that the following criteria serve for the interpretation of the photographs taken:

a) Colour: Changes in colour are the basic criterion for discerning almost every kind of affliction.
b) Crop cover: Changes in the density of crop cover show where and when growth is being inhibited.
c) Distribution: Determination of the distribution of symptoms and of successive changes in their distribution indicates whether the affliction is static or dynamic, and whether or not it is randomly distributed, which is most frequently due to pests or diseases (see Chap. 2.2.2).

2.4.5.4 Contribution of Aerial Remote Sensing to the Monitoring of Crops

Remote sensing by aircraft is not meant to replace the monitoring of crops from the ground, but to guide it and make it more effective and economical. The potential contributions of remote sensing in this field are:

a) Rapid coverage of large areas.
b) Early detection of phenomena that require investigation (trouble spots).
c) Assessment of the size of areas affected.
d) Annual documentation of the state of crop health, to determine changes occurring over the years.
e) Economies in the time and manpower expended on ground monitoring.

As regards the early detection of pests and diseases, it must be taken into account that this refers primarily to damage caused in the upper parts of crops: infections or infestations restricted in their early stages to lower parts of plants may be masked and undetectable from above.

One of the first advisory services to develop and use aerial infrared photography for crop disease monitoring is England's Agricultural Development and Advisory Service (ADAS), and this will be referred to in Chapter 3.1.6.

2.4.5.5 Examples of the Application of Aerial Remote Sensing in Crop Protection Work

Sooty Mould on Citrus and Cotton Crops. Sooty mould growing on the honey dew secreted by aphids and other insects cover the foliage and reduce reflectance in the 0.3 – 1.35 μm range (Gausman and Hart 1974). Early detection of sooty mould has led to prompt detection of foci of aphids (Hart and Myers 1968). Repeated aerial monitoring shows the rate at which aphids and sooty mould spread, and guides regional control operations. We have found that on citrus a scale of 1:40,000 could be used, so that each photograph covers 5000 ha (Fig. 4). On cotton, determination of plots affected by sooty mould has also proved useful in guiding the picking operation, so that dirty and clean cotton were not picked at the same time (Fig. 5).

Disease, Mineral Deficiency and Stress Assessment in Plantations. In Florida, consistent annual losses of citrus trees (primarily Valencia orange budded on Rough Lemon) are attributed to a wilting type of disease (blight), the cause of which has not been determined. It differs from tristeza virus in interfering with water uptake, while tristeza-diseased trees take up water readily. Aerial colour photography has made it possible to assess damage by recording the total tree spaces and producing lists of the productive and non-productive numbers of trees from a single photographic mission. Such photography can be used to pinpoint stress symptoms, e.g., those due to iron deficiency, nematode damage, lack or excess of water, and also weed infestations (Blazquez and Horn 1980).

Fig. 4. Sooty mould loci in citrus groves, indicated in this black and white reproduction of an infrared colour photograph by *black patches*

Fig. 5. Sooty mould on cotton (*below* and *on the right*) as a result of infestation by the tobacco white fly (*Bemisia tabaci*). The photograph was taken after the field had been sprayed repeatedly by aircraft, which was, however, prevented by obstacles from covering the edges of the field, and this is where the infestation was found to develop

Fig. 6. Avocado grove in which one part (*right*) has been severely affected by chlorosis. (Black and white reproduction of a colour infrared photograph)

Deficiencies of mineral elements have shown up clearly in photographs of avocado plantations in Israel. Four grades of iron deficiency could be distinguished, and this served to guide the grower in deciding on the dose of iron to be applied to correct the deficiency (Fig. 6) (Hochberg and Edelbaum, unpubl.).

***Sclerotinia* Blight of Peanuts.** The severity of blight (*Sclerotinia sclerotiorum*) on peanuts, as detected by aerial infrared photography, was correlated with actual yield losses in the field. Peanut pod losses in the soil increased as disease severity, determined by imagery interpretation, became more pronounced. In areas of fields interpreted to be slightly, moderately or severely infected with the blight, reduction of pod yields due to infection were, respectively, 2, 5, and 7 times greater than yield reductions from other causes normally expected in areas not affected by blight. Sclerotial populations were ten times greater in soils designated as severely infested than in soils from slightly or non-infested areas (Porter et al. 1977).

Late Blight of Potatoes (*Phytophthora infestans*). Since late blight is favoured by the wet conditions in low areas of the field or near ditches, initial surveys on potatoes in Florida were made to determine the location of such areas, followed by flights every 3 days to monitor initial infection and subsequent spread of the disease. Loci of infection were in fact found near wet spots in two areas, and continued aerial surveillance was used as a guide for fungicide application to the field. A secondary benefit of the photography was the location of an area in the field that would not maintain water at the proper depth for growing po-

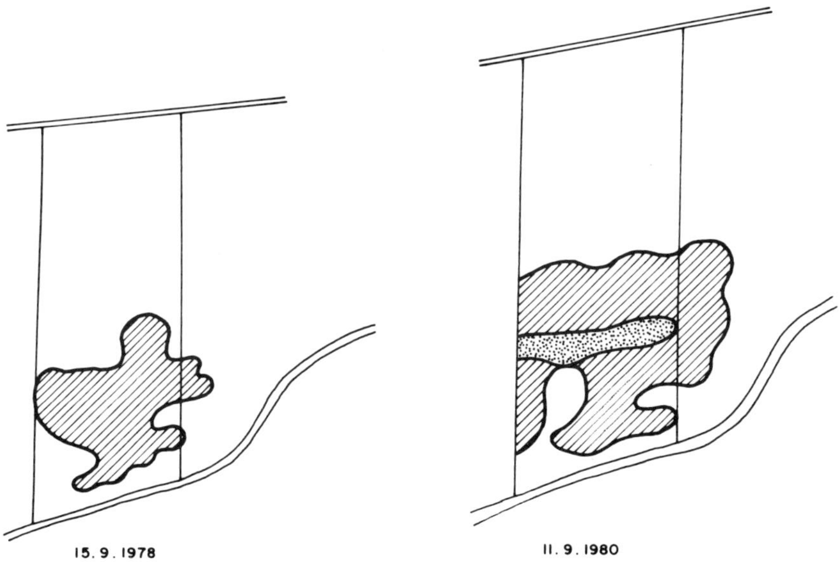

Fig. 7. Spread of *Meloidogyne incognita* infestation in a cotton field from 1978 to 1980, as seen in maps drawn from colour infrared photographs. *Dotted area* indicates the highly infested area of the field

tatoes, because it was a natural sinkhole filled with sand, allowing irrigation water to percolate deeply beyond plant roots (Blazquez et al. 1981).

Nematodes in Cotton. Surveys of cotton fields in Israel's Hule valley were made over three years to detect areas infected with the root-knot nematode *Meloidogyne incognita* and monitor the spread of infection. On the colour infrared photographs, nematode infestation appeared as irregular areas characterized by exposed soil and dense weed vegetation. The centre of a highly nematode-infested area was characterized by lack of vegetation or by thin rows of cotton, which could also be seen along the margins of the area. Photographs taken at the end of the cotton-growing season gave a better image of the infested area than photographs taken earlier in the season. The rate of nematode spread according to the photographs averaged 25–50% per year (Orion et al. 1982) (Fig. 7).

References

Blazquez CH, Horn FW Jr (1980) Infrared photography: application in citriculture. NASA Ref Publ 1067, 82 pp

Blazquez CH, Elliott HA, Edwards GJ (1981) Vegetable crop management with remote sensing. Photogramm Eng Remote Sensing 47:543–547

Gates DM (1970) Physical and physiological properties of plants, remote sensing with special reference to agriculture and forestry. Natl Acad Sci, Washington DC, pp 224–252

Gausman HW (1974) Leaf reflectance of near infrared. Photogramm Eng Remote Sensing 40:183–191

Gausman HW, Hart WG (1974) Reflectance of four levels of sooty mold deposits produced from the honeydew of three insect species. J Rio Grande Valley Hortic Soc 28:131–136

Hart WG, Myers VI (1968) Infrared aerial color photography for detection of brown soft scale in citrus groves. J Econ Entomol 61:617–624

Knipling EB (1970) Physical and physiological bases for differences in reflectance of visible and near infrared radiation from vegetation. Remote Sensing Environ 1:155–159

Orion D, Nessim-Bistritsky B, Hochberg R (1982) Using color infrared aerial photography to study cotton fields infested with *Meloidogyne incognita*. Plant Dis 66:105–108

Porter DM, Powell NL, Cobb PR (1977) Severity of Sclerotinia blight of peanuts as detected by infrared aerial photography. Peanut Sci 4:75–77

2.4.6 Optimization of Pest Management Advice by Computer-Based Decision Making

2.4.6.1 Computerized Systems for Pest Management

R. AUSHER and J. PALTI

Adoption of the Computerized Advisory Approach

The adoption of computer-supported approaches to advisory work results from the increasing need of intensive agriculture for quantitative information; this constitutes the basis for taking full advantage of the biological potential of crops and livestock and of the managerial potential of advanced farming techniques. Recent developments in electronic technology make it possible to store and retrieve data efficiently and to analyze complex production and crop protection functions and their possible interactions in *real time*.

The production and adoption of the computerized advisory approach depends on three principal elements: the human factor, software, and hardware.

The human factor is a decisive component in the computer-oriented approach to all professional tasks, and this is most certainly true where advancement of the farming community is concerned. Programmes are usually developed by keen computer-minded advisers, and this is to be welcomed for setting patterns for subsequent use by advisers. Programme development should focus on the complex information needed at farm level, should proceed in cooperation with farmers and their representatives, and should then, where necessary, be worked out in cooperation with research. Adoption of computer support in decision making by advisory personnel should begin with a series of orientation activities to familiarize first managerial and later field level personnel with the computerized approach. Training activities, at first introductory in nature and then programme-specific workshops, should make the personnel familiar with the programmes to be used, and attempt to raise them to a certain level of proficiency in programming and system analysis.

Software and programme development have to be the outcome of the joint definition of actual regional needs. They will thus be user-oriented and provide answers to real problems and requirements.

Hardware is here considered last, as it should be adapted to serving the storage, computing, and dissemination needs of the programmes drawn up.

The above considerations are valid for advisory work in general, and their application to pest management will be discussed in the following sections.

The Development of Computerized Monitoring and Information Delivery for Pest Management

Over the past 20 years, the use of computer-based decision-making in crop protection has advanced most rapidly in the United States. These advances have recently been reviewed by Welch (1984), and readers wishing to get acquainted with this subject in greater detail are referred to this review. We shall here briefly describe some examples of the successful use of computer schemes in advisory work, with special reference to pest management. The schemes now in use include crop production models, data bases, as well as information delivery systems.

In the United States the close link between universities and advisory services has proved highly productive for the development of such schemes. The first computer-based information delivery system focused on pest management went into operation at Michigan State University in 1975 (Croft et al. 1976). It was called the Pest Management Executive (PMEX) system, began as an apple pest management program, but was rapidly expanded to pests of field and vegetable crops. In the course of a few years this system developed into the Pest Event Scheduling System described more fully later in this section. In general, however, development of pest management schemes during the past 10 years lagged behind the formulation of such schemes in areas like budgeting, accounting, record keeping, linear programming, etc., objectives that are well-defined and use relatively simple algorithms. By contrast, in pest management, algorithms are complex and require intensive research.

Nevertheless, the progress made in programme development has been significant, and the range of information provided has widened steadily. Thus, in the realm of data bases, the National Pesticide Information Retrieval System (NPIRS) developed at Purdue University (Cress 1983) stores data on pesticides and their label information. In other dynamic systems, such as the Dutch EPIPRE (Zadoks 1981), information on costs of inputs and market prices are added to the relevant pest observations (see below).

Important advances have been made in the delivery of computerized advice to extension staff and growers. This is most striking where users have direct access to terminals or possess terminals of their own. Among the major computerized information delivery systems developed in the United States, mention should be made of (a) the System for Computer-Aided Management of Pests (SCAMP), which links all the 12 north-eastern states in the U.S. (University of New Hampshire 1983); (b) the AGNET system providing information to widely dispersed, diverse crop protection users in 45 states (Nieuwsma 1984); (c) the TELFARM and TELPLAN systems in Michigan; (d) the FACTS system in Indiana (FACTS 1980); and (e) the BUGNET and other systems developed in Texas and described in some detail in Chapter 3.6.4.

A completely different approach is that of using computer programmes and microprocessors that can be placed into fields and are dedicated to giving field-specific forecasts for individual diseases. For the purpose of guiding the control of apple scab, *Venturia inaequalis*, such a scheme was first proposed in Germany by Richter et al. (1973), and has since been further developed in Michigan, U.S.A. (Jones et al. 1984); a similar scheme relating to potato leaf blight (*Phytophthora infestans*) has been described in Section 2.4.2.3. Microprocessors that will stand up to field conditions (dust, moisture, chemicals, noise from farm machinery) have been developed (Jones et al. 1984). The disadvantage in the development and use of this approach is that it is incompatible with crop production models or general information delivery systems.

In most countries in which computer-based pest control advice has become available, it is still disseminated chiefly through crop protection bureaux or growers' organizations, such as farmers' cooperatives.

In Israel, where many farm communities are collective settlements (kibbutzim), these have been quick to initiate computer services in the framework of regional enterprises; computer terminals located at each settlement readily lend themselves to pest control work, but the growers have to compete for computer time with the administrative (accounting) and industrial branches of the settlement. The regional advisory bureaux have computer links with the regional enterprises, so that rapid transmission of urgent messages becomes easy (see also Chap. 2.4.6).

To illustrate how computer-based systems work in pest management under diverse conditions, we have selected four operative schemes. These include one from the United States where up-to-date information on additional schemes can readily be obtained (Jones et al. 1984, Welch 1984); and the BUGNET programme operative in Texas as described in Chapter 3.6. The second scheme selected is the EPIPRE system, which has been developed in The Netherlands and is being adopted in several countries in western Europe (Zadoks 1981, 1984). Two forecasting and monitoring schemes from the German Democratic Republic are then presented as examples of schemes operative in eastern Europe, which have not so far been reported extensively in international publications (Gutsche and Kluge 1981, Herold and Sachs 1982).

This is followed by two Israeli schemes developed by or in close cooperation with the advisory service. The cotton scheme (Sect. 2.4.6.2) stresses the importance of converting pest population levels, as determined by field counts, into damage levels, taking into account food preference and life expectancy of larvae. The wheat scheme (Sect. 2.4.6.3) presents thresholds worked out not only for the initiation of control measures but, no less important, for their timely discontinuation.

Examples of Computerized Pest and Disease Management Schemes in Operation

a) The Pest Event Scheduling System in Michigan, U.S.A.

The Pest Event Scheduling System for Biological Monitoring and Pest Management has been developed in Michigan and has been described by Gage et al. (1982). The components of this system are: (1) A computer-based weather information management and delivery system to provide daily weather data and temperature forecasts based on 61 weather stations in or near Michigan. (2) A data base (code-named PESTAPP), which contains observations on major pests, their life stages, degree-day occurrence range, and hosts. This, in conjunction with the weather data, is used to estimate when a pest life stage is or will be occurring at any particular Michigan location. (3) A predictive extension timing estimation system (PETE) to estimate the phenology of a more limited number of pests on the basis of carefully researched parameters affecting each stage in their life-cycle (Welch et al. 1978). (4) Combinations of current weather data and the data provided by PESTAPP or PETE in a pest-scheduling system (BIOSCHED) designed to enable the observer in the field to estimate pest density and take pest-management decisions. The observer must be able to identify the pest correctly and to sample the populations reliably. BIOSCHED will assist him by indicating what pests are likely to occur in his area at that particular time, and what stage of their life cycle should be looked for. To further assist the observer, a cooperative crop monitoring system (CCMS) has been developed, in which observations on crops and their pests are recorded, stored in computers, and easily retrievable.

In practice, this works as follows: an adviser or grower gets in touch with BIOSCHED via remote terminal and requests a listing of pests by location and crop. He is provided with pest names and their current developmental stage and host, and with threshold temperatures, degree-day accumulation, and 7-day predictions. He then requests from CCMS the pest severity level, sorted by county and crop, and the specific location of each field or orchard, where a given pest has been recorded in numbers exceeding a specified threshold during the previous time interval. All this information serves to take pest control decisions or to guide monitoring work to the most appropriate period. If necessary, supplemental regional information is also available from state pest management specialists through a computer-based pest management delivery system.

b) The Dutch EPIPRE System for Disease and Pest Management in Winter Wheat

Zadoks (1981, 1984) has described the EPIPRE system developed in the Netherlands for winter wheat production. A diagram of the scheme appears in Fig. 1. At the heart of this scheme is the data bank in which two kinds of data are stored: (1) the farmer's basic data for each field, collected in early spring and including name of cultivar and varietal resistance, sowing date, soil type, and the nitrogen status and developmental stage of the crop; (2) variable data supplied by the farmer as he monitors his crop all through the season. Participant farmers send in these data on reply cards, and if they fail to do so in time, they receive an automatic reminder. The above data are collated in the data bank with weather information and with epidemiological models for six fungal diseases and all the aphid pests of wheat.

On the basis of all the input listed above, and of additional data reflecting current costs and prices, the computer model not only issues field-specific recommendations whether or not the time has come to perform control operations, but also provides financial justification for such recommendations in terms of added value (expressed as kg ha^{-1}).

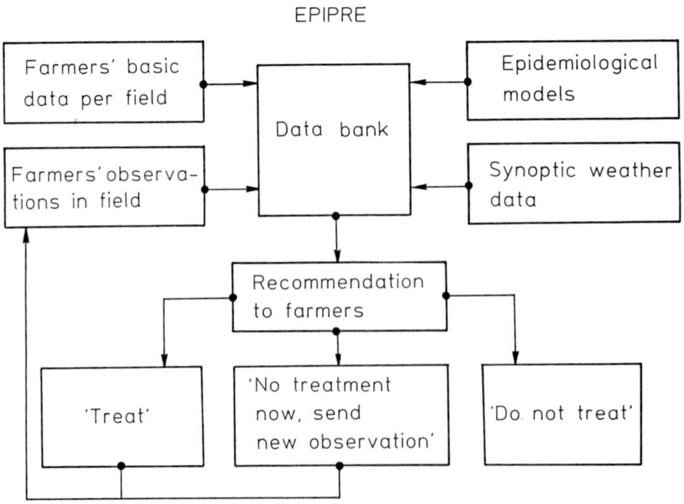

Fig. 1. Block diagram of the EPIPRE scheme (Zadoks 1981)

Participant farmers pay a fee; the rapidly growing number of participants would seem to indicate that such payment is quite profitable. The EPIPRE system is now also in operation in Belgium and Sweden, and is being tested in other West European countries.

c) PHYTEB Prognosis of *Phytophthora infestans* on Potatoes in the German Democratic Republic (G.D.R.)

Reference has been made in the preceding chapters (Chaps. 2.7.1 and 2.7.3) to the PHYTEB prognosis scheme for the prediction of potato blight. Two computer models are used to forecast blight outbreaks in each of the 13 zones into which the G.D.R. has been divided for this purpose (Gutsche and Kluge 1983).

The SYMPHYT I model serves to predict, chiefly on the basis of blight incidence in the preceding autumn season, the date before which blight is not expected to appear, i.e., it issues a negative prognosis. For example, since blight infection of seed potatoes harvested in autumn 1982 was low, the disease was predicted not to appear anywhere in the G.D.R. before the third decade of June 1983.

The SYMPHYT II model deals with the prediction of the first seasonal appearance of potato blight and with the subsequent course of the disease. Beginning on May 20th, calculations are made three times a week. To begin with, the amount of inoculum (infection pressure) is estimated according to the amount of inoculum deriving from seed potatoes, with due consideration of the actual rainfall in the month of June. If this does not indicate conditions supporting blight outbreak, a negative prognosis is issued for the following 7–10 days. But if inoculum and weather conditions appear favourable to blight, a much more detailed forecast is worked out, according to (1) varietal susceptibility and the crop's stage of development; (2) specific weather conditions in the zone; and (3) cultural conditions (irrigation, date of sowing etc.) and other regional parameters; fields at altitudes of more than 500 m above sea level are excluded from the scheme.

Under the weather conditions prevailing in the G.D.R., a sequence of dry days not infrequently disrupts or ends blight development. After dry periods exceeding 7 consecutive days, growers are advised to discontinue blight control treatments, provided that previous blight development in the field has not affected more than 10% of the foliage, and that neighboring fields are not blighted severely.

The PHYTEB scheme is centrally operated by the Meteorological Service. Forecasts are transmitted by telex to the plant protection bureaux in each zone.

d) Computer-based Monitoring of Cereal Pests and Diseases in the German Democratic Republic (G.D.R.)

The scheme operative in the G.D.R. for the monitoring and management of cereal pests and diseases (Herold and Sachs 1982) consists of two parts. (1) The pest and pathogen monitoring service (Schaderregerüberwachung) which is centralized and run by the state's crop protection authority. This service collects annual data from all regional crop protection bureaux; in each region, 30 fields of each major crop are selected at random and are monitored throughout the season. The data are transmitted to a central computer, which informs all regions, within 48 h, of the distribution and incidence of the various pests or diseases. Seven cereal diseases and four pests are at present covered by this scheme, and seven major weeds have recently been added. The aim of this service is to use the data collected over many years to predict the likelihood of pest and pathogen populations reaching levels apt to cause commercial loss in the current year's crop. (2) Field-specific monitoring (Bestandsüberwachung), carried out by crop producers, which is the basis for operative pest control decisions. Such decisions are based in part on the centralized pest and pathogen monitoring service, which indicates when and at what intervals to inspect the field. But decisions whether or not to execute control operations in a given field rely primarily on specific field data: stage of crop development, incidence of pest and disease in relation to the economic thresholds worked out for each of them, parasite and predator populations, infestation of neighbouring fields, and local weather and weather forecasts.

Since the price levels of grains in the G.D.R. are fixed by the state, considerations of variations in crop value are not required to define treatment thresholds.

How Can Advisory Personnel Promote Development of Computerized Pest Management Schemes and Their Adoption by Growers?

a) Development and Updating of Schemes. The data on which pest management models have to be based derive from many years' field observations and trials combined with the findings of laboratory or climate chamber research. The models have often been developed by research personnel, and the weakness of such models may lie in the fact that neither advisers nor growers have been involved at an early stage. Implementation of such models then frequently encounters obstacles that are difficult and costly to overcome.

In Israel, crop protection models for cotton, potato, and groundnut have been initiated by the Extension Service. Development of these models has often been preceded by operation of crop-monitoring programmes in which growers cooperated. More sophisticated models, developed subsequently in collaboration with research institutions, have been based on the earlier programmes, which all include a crop protection component.

Once the model has been introduced into practice, whether it originates from research or other sources, its success over the years depends greatly on the efficiency with which it is continuously updated according to agricultural developments: changes in cultural conditions, e.g., trickle irrigation, new crop cultivars, changes in pesticides, and the cost of application, etc. Such updating is primarily the job of the crop protection adviser. Where models relating to one specific pest are in use, they can sometimes be adapted to additional pests, provided adequate field data are available. Updating computer schemes may, in fact, become one of the major tasks in the adviser's work.

b) Usage by Advisers. The extent to which the AGNET scheme is used by advisers in North Dakota has been analyzed by Nieuwsma (1984), who divided them into low, medium, and high user groups. Younger advisers and those with fewer years of service tended to be higher users. Groups that exert influence on the adviser, especially farmers and government agencies, were effective in encouraging the use of AGNET. Previous computer training increased use of the scheme by the advisers.

c) Usage by Growers. Introduction of computer schemes into pest management is perhaps still too recent to formulate recommendations how best to adapt them to maximize the benefits farmers can derive from them. An interesting survey of factors affecting growers' usage of such schemes has been carried out in the Netherlands by Blokker (1983) in relation to the EPIPRE system of pest and disease control in wheat. Among the points he makes we would like to quote the following as significant for advisory work:

- The majority of growers using EPIPRE have increased their ability to diagnose wheat pests and diseases and to do so at the right time. Some growers even feel that they have by now learned enough to manage on their own, and have discontinued participation in the scheme.
- Many growers would like to see the pest control programme expanded into a scheme for comprehensive crop management.
- It is not enough to issue recommendations when to spray; the growers wish to know the reasons for such recommendations, so that they can themselves participate in the decision making.

Outlook

At the present time the extent to which computerized pest management schemes are put to practical use does not yet seem commensurate with the investment made in research. In the United States, large-scale adoption of such schemes may be expected in the near future, as commercial firms move into the pest information delivery field at an accelerating pace (Welch 1984). A most promising development is the availability, on a massive scale, of inexpensive microcomputers. These facilitate constant, rapid supply of relevant data to advisers and growers, and the equipment is so versatile that it can be used for numerous other purposes.

As stated above, use of in-field microcomputers may hold the promise of tailor-made, field-specific pest management in climates and agrosystems in which regional forecasting has proved impracticable. The rate at which electronic equipment will come into use for this purpose depends primarily on the cost at which it can be acquired, the ease with which it can be made fool-proof and simple to operate, and the effort made to train advisers and growers by competent advice. It is too early to predict how soon and where this will benefit farmers in technically less advanced countries.

The progress of computer-based decision making will have an impact in many technical and structural areas of advisory work in agriculture:

a) Technical tools are developed on the basis of quantitative parameters that can continuously be updated, e.g., the frequency and dynamics of pest populations, and this will improve decision making at field level in both developed and developing agricultural economies.
b) New and improved extension methodologies are being developed for the dissemination of farm advice, including urgent warnings of pest and disease outbreaks.
c) The structure and personnel of advisory organization has to be adapted to cope efficiently with the working methods that derive from computerized decision making, e.g., the integration of pest and disease control in the framework of general crop production schemes.

References

Blokker K (1983) Computer-aided extension: the case of plant protection for wheat in the Netherlands. Paper presented at 6th Eur Semin Extension, Lucca, Italy

Cress DC (1983) NPIRS News, Purdue Univ, West Lafayette 1(2):1–2

Croft BA, Howes JL, Welch SM (1976) A computer-based, extension pest management delivery system. Environ Entomol 5:20–34

Eisensmith SP, Loria R, Olson BD, Jones AL (1980) EPISTAR: an epidemiological information storage, retrieval, and analysis system. Plant Dis 64:646–651

FACTS (1980) FACTS program planners guide. Ind Coop Ext Serv, Purdue Univ, West Lafayette, 27 pp

Gage SH, Whalon ME, Miller DJ (1982) Pest event scheduling system for biological monitoring and pest management. Environ Entomol 11:1127–1133

Gutsche V, Kluge E (1983) Phyteb-Prognose, ein neues Verfahren zur Prognose des Krautfäule-Auftretens (*Phytophthora infestans* Mont. de Bary) Nachrichtenbl Pflanzenschutz DDR 37:45–48

Herold H, Sachs E (1982) Erfahrungen mit der EDV-Schaderregerüberwachung in den vergangenen 5 Jahren, erläutert am Beispiel wichtiger Getreideschaderreger. Wiss Beitr Martin-Luther Univ Halle-Wittenberg 1982, pp 532–544

Jones AL, Fisher PD, Seem RC, Kroon JC, Van Demotter PJ (1984) Development and commercialization of an in-field microcomputer delivery system for weather-driven predictive models. Plant Dis 68:458–463

Nieuwsma L (1984) AGNET: Increasing its effectiveness. J Ext 2:11–15

Richter J, Steiner H, Schipke W (1973) Ein elektronisches Schorfwarngerät im Obstbau – Arbeitsweise und Aussichten für die Zukunft. Mitt Biol Bundesanst Berlin 151:281–282

Univ New Hampshire (1983) SCAMP on-line newsletter. UNH Comput Serv 9:15–16

Welch SM (1984) Developments in computer-based IPM extension delivery systems. Annu Rev Entomol 29:359–381

Welch SM, Crofty BA, Brunner JF, Michels MF (1978) PETE; an extension phenology modeling system for management of multi-species pest complex. Environ Entomol 7:482–494

Zadoks JC (1981) EPIPRE: a disease and pest management system for winter wheat developed in the Netherlands. EPPO Bull 11:365–369

Zadoks JC (1984) A quarter century of disease warning, 1958–1983. Plant Dis 68:352–355

2.4.6.2 Development of a Model for Monitoring *Heliothis armigera* in Cotton in Israel

Y. Sachs

In Israel, *Heliothis armigera* is the first lepidopterous pest to appear on cotton. Save for locations in which the whitefly *Bemisia tabaci* occurs, *H. armigera* is also the first pest treated with insecticide. This pest normally appears in June, shortly after the first squares (floral buds) appear, and disappears in early August.

Development of monitoring services and treatment thresholds for cotton pests has always been one of the major aims of cotton research in Israel. The early appearance of *H. armigera* and the fact that insecticides for its control greatly affect the fauna of natural enemies of cotton pests, have afforded the correct timing of *Heliothis* control marked importance in the fight against cotton pests of all kinds. But in order to determine realistic treatment thresholds for *H. armigera*, it is necessary to understand the population dynamics of this pest, and to quantify the damage it does to the cotton crop. This involves determination not only of the total number of fruiting bodies damaged, but also the distribution of damage among fruiting bodies of different sizes (Kletter and Wallach 1982). Such data have only become available recently and serve as the basis for the model to be described here briefly. But before we proceed to do this, the conventional method for monitoring *H. armigera* on cotton in Israel will be outlined.

The Conventional Procedure for Monitoring H. armigera. The conventional procedure for monitoring cotton for *H. armigera*, as described by Melamed and Shoham (1977), consists in dividing fields into reference plots measuring up to 5 ha each, and counting larvae every 3 days at five sampling points, each of which consists of a 2-m-long row section. In these sections every leaf, square and boll is checked. The eggs and larvae found are evaluated in terms of larval units, as follows:

Stage found	Larval units assigned
Eggs and/or larvae less than 4 mm long	0
Larvae 4–15 mm in length	1
Larvae exceeding 15 mm in length	3

The number of larval units have been assigned chiefly according to the relative amount of damage larvae might be expected to cause in one day. Two treatment thresholds have been defined: in the early stage of cotton growth, until two bolls of about 1 cm diameter have formed on each plant, spraying is advised when a mean of six or more larval units has been found in the five sections; but when plants have developed beyond this stage, spraying is to begin when only four or more larval units are found. However, at all growth stages, sprays are also recommended when infestation is close to, but just below, the threshold over 5–6 days, with no downward trend apparent by the 6th day.

This procedure has proved fairly satisfactory over many years. It has the advantage that growers are familiar with it. But its drawbacks are that it relates directly only to numbers of pests, not the damage they cause, that it fails to account for natural death of eggs or larvae, and has assigned arbitrary unit values to larvae of various sizes. This also finds expression in sudden and unrealistic jumps in monitoring data from levels well below to levels above thresholds, e.g., when numerous larvae simultaneously reach the transition from below to above 4 or 5 mm length.

Development of the Model for Consecutive, Interrelated Sampling. The monitoring procedure to be described here has many of the characteristics of sequential sampling (Sect. 2.4.1.1). We prefer, however, not to include it in this category, because one of the aims of sequential sampling, the restriction in the number of samples to be taken in various situations, and the resultant saving in labor, was not our aim and has not been attempted in this work.

The prerequisite for the introduction into practice of a model which could use the counts made in monitoring for decision-making was the availability of a central computer and of terminals, with suitable staff, to each participating cotton grower. In Israel's coastal plain, where the model has been developed, cotton is overwhelmingly grown by collective settlements (kibbutzim) which also use computers for other purposes, and were quite ready to use them for cotton pest control as well.

The sampling and counting procedure used for the model did not differ from the conventional one described above. However, important new aspects were introduced in the "weighting" of larvae of different size to express the damage they may be expected to cause. Account was now taken not only of the daily damage a larva of a given size can cause, but also of the number of days the larva may be expected to live and gnaw with ever-rising efficacy, and of the reduction of damage that could be secured by prompt spraying. Experiments to quantify expected damage and damage amenable to prevention by prompt and

Table 1. Expected damage by *Heliothis armigera* to cotton and expected reduction in damage due to insecticide use (Wallach, unpubl.)

Phase	Number of			
	Individual fruiting bodies expected to be damaged unless sprayed		Individual fruiting bodies expected to be saved from damage when sprayed in time	
	Squares	Bolls	Squares	Bolls
Eggs	2.3	0.1	0	0
Larvae:				
up to 4 mm	2.9	0.2	2.5	0.2
4–10 mm	3.7	0.3	3.1	0.3
10–15 mm	3.2	0.4	2.7	0.3
more than 15 mm	1.8	0.2	1.5	0.2

effective spraying, have given the results briefly summarized in Table 1 (Wallach, unpubl.). In these figures eggs, when hatching, are considered to cause appreciable damage, which is not preventable by the insecticides commonly applied to cotton in Israel. Larvae up to 15 mm in length may cause even more damage, if unchecked, but they are susceptible to spray and therefore reduction of damage by timely insecticide application is estimated at about 85%. However, fully grown larvae, due to pupate soon, have only a limited time to feed, and the damage they are expected to cause is thus lower; it can be prevented to a large extent by spraying.

On the basis of these data for larval survival and estimates of damage, the following values have been adopted and termed "larval units" (L.U.) for the purpose of weighting to fix thresholds for application of insecticides: eggs = 0 L.U., larvae of any size up to 15 mm = 2 L.U., larvae more than 15 mm in length = 1 L.U.

After the data collected by monitoring were fed into the computer model, the latter thus performed the following operations:

1. Weighting the larval counts according to expected damage and proportion of that damage that could be reduced by spraying, i.e., converting counts into weighted larval units.
2. Relating backward the figures obtained on a given date to those obtained in previous samplings. This achieves two objectives:
   a) it checks the reliability of the samples freshly taken against samples taken previously; if there are obvious and unexplainable discrepancies between consecutive samplings, a correction can be made;
   b) it may introduce pest infestation levels as a factor in making the decision whether or not to spray, and this is of special importance where levels stay near or just below threshold values for a number of days.
3. Projecting a picture of the infestation about to develop, by converting the present level into that to be expected shortly.
4. Conversion of population into damage levels not only by the weighting method described above, but also by taking into account the larvae's food preference (squares or bolls), life expectancy and mortality due to natural enemies and adverse environmental factors, such as hot, dry winds. Mortality values were derived from an experiment carried out in 1980 (Sachs and Wallach 1984). Research to obtain better quantification of this aspect is under way.
5. Assessing the extent to which the crop can compensate for squares and bolls affected by larvae; this refers to the numbers of such organs formed, and especially to the question whether the rate at which they form will enable them to ripen until the crop is picked. The capacity to compensate obviously lessens as the crop approaches maturity.
6. Determination of the number of fruiting bodies to be expected at harvest with and without insecticidal treatment, with due regard to the fact that the treatment can only be carried out on the following day.
7. Issuing recommendations to spray, if at least 60 fruiting bodies per meter row have been affected, and at least ten fruiting bodies of those not yet affected can be saved by spraying.

This model, though still in the development stage, has been successful in the region in which it has been developed and tested by numerous cotton growers. It assesses with higher accuracy the capacity of larvae to harm cotton and introduces some estimate of the rate of larval mortality due to natural causes. The accuracy of field monitoring has improved as the monitor feels "Big Brother" looking over his shoulder and does not like to be caught out in erroneous sampling and counting, and because errors in larval counts can in fact be corrected by relating figures to previous counts. A saving of 30–60% in control operations specifically directed against *H. armigera* has thus been achieved without increase in risk. First applications have usually been delayed, with the result that the period over which natural enemies of all pests remain active in the field has been extended.

This model is being developed further as part of a general cotton pest control model which, in its turn, forms part of a comprehensive cotton growing model. In future phases pests will be evaluated as one of the factors determining the growth of cotton, as interrelated with other growth factors, especially fertilization and irrigation. Recommendations for the timing of control operations should then be adapted to the specific growing conditions and pest infestations in each field.

References

Kletter E, Wallach D (1982) Effects of fruiting form removal on cotton reproductive development. Field Crop Res 5:69–84

Melamed Y, Shoham Ch (1977) Insect scouting in cotton fields. Isr Minist Agric, Ext Serv, 17 pp

Sachs Y, Wallach D (1984) *Heliothis armigera* in cotton: damage, development and mortality. Publ 31, Hadera Ext Bur, 18 pp

2.4.6.3 Optimization of the Control of Wheat Leaf Diseases by Integration of Advisory and Research Work

A. DINOOR

Introduction

Epidemics of leaf rust (*Puccinia recondita*), stem rust (*P. graminis*), yellow rust (*P. striiformis*) and Septoria leaf blotch (*Septoria tritici*) threaten wheat grown in Israel, sometimes with devastating effects. But no accurate information on actual losses and on cost–benefit relations in chemical control has so far been available. Moreover, wheat is grown as an extensive crop, giving a relatively low income, and benefits from chemical control are often marginal. However, the alternative use of resistant varieties has proved ephemeral in its reliability because of frequent breakdown of resistance.

Fungicidal control of wheat leaf diseases constitutes one of the central problems in advisory work, for several reasons:

1. Fields are large and variable, requiring large-scale inspection work, which is time-consuming and often tedious.
2. The economic benefit of control operations is not immediately apparent, may often be in doubt, and growers need expert advice to decide whether or not to perform control operations.
3. Growers fail to detect early symptoms and become concerned only when the epidemic is already under way.
4. Field experimentation is difficult since, due to great variability within fields, control effects are frequently not statistically significant.

The agricultural advisory service in Israel must be credited with provoking the interest of growers in potential improvements of control and with drawing the attention of research institutions to this problem. The advisory service promotes an annual "Operation Wheat", to encourage competition between farmers in increasing wheat yields. Production factors such as the application of water and fertilizers and the control of weeds and diseases have been utilized by competing growers to the fullest extent, sometimes even beyond the bounds of economic reason. In the course of this operation, farmers came to realize, among other things, how effective disease control could be. But no cost-benefit calculations had as yet been made, and advisers were not overly impressed by the fact that the fields looked healthier for a longer time: an economic analysis was urgently needed, and applied field research was evidently called for, with strict definition of practical objectives.

Objectives of Research

The first problem to be tackled was the timing of fungicidal applications, limited in number as they are by their fairly high cost relative to the value of the additional yield they may produce. The cost/benefit ratio in itself is subject to changes in various crop situations, and as the prices of fungicides and of wheat change.

The main issue was therefore how to reduce the number of applications. It soon became evident that lengthening the intervals between them cannot lead to a reduction in the number of applications; the intervals are determined by the persistence of fungicidal deposit and by the growth rate of the wheat. They cannot therefore be lengthened unless more persistent fungicides become available. It was left to research to determine, how to shorten the period over which applications have to be made, or to find more persistent fungicides.

Thresholds for Initiation of Control

The period of control could be shortened by delaying the first application, after determination of a realistic threshold for applying the first spray. This involves experimentation to indicate control thresholds that can be ascertained visually without too much difficulty. The thresholds to be tested were adapted to the disease in question.

For leaf rust, which spreads by wind, the control threshold was studied experimentally and was fixed at the low level of 100 pustules per sample of 100 tillers. But it was found that in actual practice this threshold is too risky because of the time lag between its determination and the performance of the control operation: by the time the field was in fact sprayed, the disease had advanced beyond the postulated threshold level. In order to be of practical value, the threshold value thus had to be lowered. It was eventually fixed, for leaf rust control by protectant fungicides, at 50 pustules per random samples of 100 tillers.

Septoria leaf blotch is dispersed by the splashing of inoculum from lower to upper leaves. The uppermost leaf, the flag leaf, is the most important in determining yield, and the later the disease reaches this leaf, the smaller the damage it causes. The thresholds to be tested for leaf blotch control were thus the proportion of diseased leaves at different heights, counting the leaves back from the top leaf. Experimentation showed that the threshold for initiation of control could be fixed at 50% disease incidence on the fourth leaf from the top; this top leaf is usually the flag leaf, but if the disease develops early, the top leaf at the time of sampling can be the leaf which will eventually be the one below the flag leaf.

These thresholds for leaf rust and leaf blotch, determined at first for control by protectant fungicides, were maintained at the same level for systemic fungicides, despite the fact that the latter can eradicate both fungi up to 7 days after penetration; a desirable margin of safety was thus secured.

To enable farmers or inspection personnel ("wheat scouts") to ascertain thresholds for the initiation of control, they should be trained in correct sampling procedures and in the recognition of disease symptoms. Annual training meetings have been held by the advisory service as shortly as possible before the critical season. Colour slides, as well as diseased plants, were used for demonstration, and field exercises were conducted.

In the course of the season, growers and wheat scouts report their observations to the regional extension bureau. Here the data are assembled and weekly bulletins are issued to indicate the state of disease in the region's fields.

Thresholds for Termination of Control

At a somewhat later stage of this work it was realized that an additional way for shortening the period of control applications could be termination of the latter well before the end of the season. This was studied by comparing effects of 1, 2, 3, or 4 applications, beginning after the initiation threshold had been reached, to delay the onset of disease as long as possible. It was found, for both leaf rust and leaf blotch, that delaying disease development beyond the milk stage failed to increase yields (Dinoor et al. 1976, Dinoor 1980, Dinoor and Shtienberg 1983). In fact, late treatments tended to decrease yield, although the

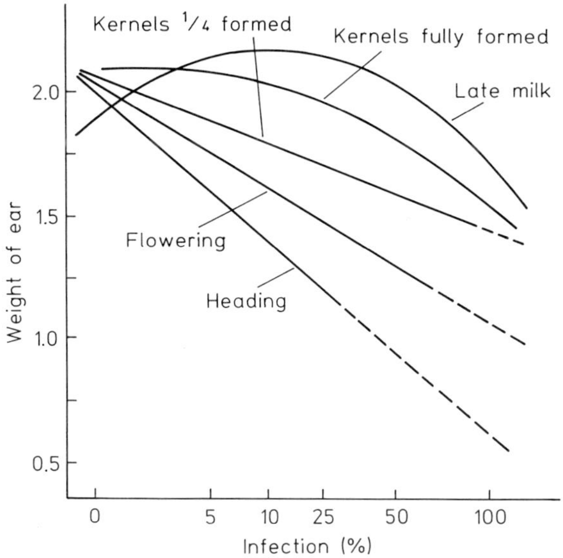

Fig. 1. Effect of percentage infection of wheat by *Septoria* leaf blotch on ear weight at different stages of plant development (Shtienberg et al. 1984)

decrease was in most cases not significant. Needless to say, such unnecessary application reduced the net income.

When yield was related to disease severity at the milk stage, the relationship was not linear, but parabolic. This, in effect, means that growers should cease spraying not later than the late milk stage. Further experimentation will have to show whether spraying should discontinue even earlier. Meanwhile, a much more detailed study with single tillers showed that the relation of yield to disease changes with the age of the infected plant (Fig. 1). When disease severity is estimated at the stages of tillering, flowering or grain filling, the relationship is linear, but at later growth stages it becomes parabolic. These findings may be taken to indicate that spraying against leaf rust should cease about 2 weeks before late milk stage, and spraying against leaf blotch about 18 days before that stage (Shtienberg et al. 1984). However, these conclusions remain to be verified experimentally and may well depend on the level of disease reached at the milk stage.

Although experimental evidence shows that spraying beyond the late milk stage fails to increase yields, it is not easy to convince growers that they should stop spraying while disease is still rampant in their fields. Advisory staff is here faced with a difficult task of explaining and persuading.

The Role Played by Climatic Conditions and the Decision-Making Scheme

The above conclusions, and the recommendations based on them, have been arrived at after study of disease development under normally humid conditions. Development of *Septoria* leaf blotch depends on rain or overhead irrigation, which may also favour leaf rust development, but to a lesser extent. Absence of rain or refraining from sprinkler irrigation may help to reduce the incidence of leaf blotch appreciably. In that case fungicide applications may or may not be redundant, and the grower needs competent advice. This can only be given when treatment thresholds and weather conditions are duly taken into account.

In order to facilitate decision-making with regard to control of leaf blotch, a scheme has been published in form of a flow-chart, which can be followed by growers as well as advisers (Shtienberg and Dinoor 1983). The key factors of this scheme are:

1. The threshold for control by fungicides is disease incidence of 50% in the fourth leaves (counted from the top).
2. Rain or overhead irrigation amounting to at least 15 mm are considered to promote disease.
3. Control operations are terminated at the late milk stage.
4. The incubation period for leaf blotch is assumed to last not less than 14 days.
5. Use is made of systemic fungicides which can eradicate the fungus up to 7 days after penetration and act for 3 weeks.

The Interplay of Economic and Biological Factors

Economic feasibility of fungicidal control of wheat diseases is conveniently calculated by translating the cost of control into units of yield. As mentioned above, this is subject to marked changes: when the study was begun, the cost per ha of a single treatment (fungicide and application) was equivalent to about 50 kg of wheat. However, with the introduction of new fungicides and owing to the general rise in prices, the cost of a treatment per ha has now risen to the equivalent of about 200 kg of wheat. Average yield loss from either leaf rust or leaf blotch in untreated fields comes to 8 – 10%. This means that fields in which yields of more than 2000 kg/ha^{-1} may be expected pay for one application per season, while fields in which yields can be expected to reach at least 4000 kg/ha^{-1} will be able to pay for two applications.

The basic concepts of loss and benefit should be considered at this point. Any increase in net income is undoubtedly the principal criterion to justify fungicidal control. However, in our opinion, consideration of this criterion alone does not constitute the whole rationale for control treatments, for two reasons: even if the yield increase secured by fungicidal treatment does no more than cover the cost of the treatment, the latter has in fact not cost the grower anything, and it has "insured" him against yield reduction he might have incurred if the weather had turned unexpectedly favourable to the disease. However, the treatment may lead to a cash flow problem: since the cost of treatments usually has to be defrayed promptly, but growers are paid for their yield several months later, a financial problem may arise which could detract from the benefits of the treatment.

Computerized Advisory Schemes in Aid of Decision-Making

Many factors affecting the grower's field and his pocket have to be integrated to arrive at a sound decision regarding control treatments and their economic value. Since each field is an individual "patient", the calculations involved are numerous and complicated. It would, however, be simple to integrate all necessary data in a computer programme to come up with meaningful advice, especially if simple estimates of expected losses can be included in the data provided. Programmes of this kind work successfully in various countries abroad. In Israel, especially in the communal settlements (kibbutzim) in which regional computers and home terminals are widely used for farm management, there should be little difficulty in evolving a practical interactive programme for wheat diseases. Such a programme is now being implemented.

References

Dinoor A (1980) Control of *Septoria* leaf blotch in wheat by means other than disease resistance. Proc 5th Congr Medit Phytopathol Un, p 174

Dinoor A, Shtienberg D (1983) Optimization of chemical control and yield loss assessment of leaf rust and *Septoria tritici* in wheat. Proc 4th Int Congr Plant Pathol, p 214

Dinoor A, Prusky D, Biali M, Neubauer J (1976) Chemical control of wheat leaf rust: a regional approach. Proc 4th Eur Medit Cereal Rusts Conf, pp 96–97

Shtienberg D, Dinoor A (1983) Optimization of chemical control of *Septoria* leaf blotch in wheat. Hassadeh 64:190–195 (in Hebrew)

Shtienberg D, Dinoor A, Marani A (1984) The relation between the incidence of wheat leaf diseases and wheat yield in Israel. Hassadeh 65:434–438 (in Hebrew)

2.4.7 Regional and Communal Considerations in Pest and Disease Control Advice

J. PALTI and R. AUSHER

Pest and disease control is essentially a regional task, and control of crop parasites is increasingly becoming a communal concern (Palti 1981). The crop protection adviser is in a unique position to view pest and disease problems from regional aspects. Some of these are described in the following:

2.4.7.1 Regional Warning Systems and Detection of Trouble Spots

The early warning, trapping and monitoring work described in some of the foregoing chapters has to be fitted into a regional scheme for the timing of control operations on individual farms or on a regional or communal scale. After a few years' observations, it is often possible to pinpoint "trouble spots" in which parasites tend to develop early and severely, e.g., on low-lying ground, near hedges, or simply in orchards that are old or neglected. Monitoring such trouble spots can save a great deal of time which otherwise would have to be spent in routine work, and aids in the crystallization of regional warnings. The regional approach to warning, before attack of pests and diseases are due to affect crops, has made much progress with the introduction of computer-based management schemes (see Sect. 2.4.6.1).

The adviser also has to keep abreast of pest- and disease-related developments in neighbouring regions. This refers in particular to (a) the time of appearance of parasites and the danger of pests migrating or spores being blown in from outside regions, and (b) development of insect or fungus populations developing resistance to pesticides. As far as interregional spread of disease is concerned, it is sometimes possible to establish interregional sequences of parasite appearance. Thus, in Israel, the late blight of potatoes and tomatoes (*Phytophthora infestans*) will not appear in the Jordan Valley unless crops in the coastal plain have previously been infected. The adviser in the Jordan Valley therefore keeps a close watch on the blight situation at the coast, because appearance of the blight there tells him when to monitor crops in the valley.

2.4.7.2 Sanitation and Detection of Parasite Sources

Sanitation in all its aspects must be a central consideration in the crop protection adviser's work. This includes crop residues and all sorts of agricultural waste, as well as water, as sources of pest or disease, and transmission of parasites from crop to crop or from wild vegetation and weeds to crops.

Agricultural Waste and Water. How does the farmer dispose of crop waste? In many cases he incorporates it in compost or leaves it in heaps near sorting and packing sheds. Many diseases, especially those forming sclerotia, survive in compost, and when this is transported to other fields it is also a source of disease to fields en route. Potato clamps are notorious for harbouring tubers affected by late blight (*Phytophthora infestans*) and as the source of blight in the subsequent season, and heaps of discarded onions were similarly found to be primary sources of mildew (*Peronospora destructor*) in Poland (Rondomanski 1969). Fruit tree prunings frequently harbour various stages of beetles and other pests. But even more important may be the transmission of parasites by water: streams running through agricultural land have been shown to carry a great variety of all types of crop parasite; drainage water, whether from rainfed or irrigated crop land, as well as irrigation ponds and channels, carry many pathogens, especially agents of soil-borne diseases and nematodes (Palti 1981). It is up to the crop protection adviser to detect such sources of trouble in his region, and this often requires some ingenuity and much tenacity.

Parasite Transmission by Weeds. Weeds are not restricted to crop land, and neither are their multihost pests and diseases, especially numerous highly mobile insects, vector-borne virus diseases and dioecious fungi, which develop part of their life-cycle on weed hosts, e.g. rusts of stone fruit trees which have species of *Anemone* as alternate hosts. But weeds on non-crop land are often considered "nobody's business", unless their potential danger is constantly brought to the attention of the farming community by a competent observer, which could and should be the crop protection adviser.

Transmission from Minor to Major Crops. Many instances are known of crops of minor importance, or even flowers or trees grown here and there in backyards, transmitting pests and diseases to major crops. Well-known is the case of the Western Yellows and Beet Yellows viruses of the major sugarbeet crop in the Pacific North-West of the U.S.A. being transmitted by the aphid *Myzus persicae* overwintering on occasional peach trees (Henneberry and George 1969). Two observations from Israel may be mentioned: (a) the Egyptian cotton worm (*Spodoptera littoralis*) in the Jordan Valley passes the winter in the unimportant fodder beet crop and then causes early attacks on the major cotton crop; and (b) the kumquat citrus variety, grown in backyards, was found to be a major source of tristeza, the destructive virus disease that attacks most important citrus crops. The role played by the minor hosts in all the above cases was detected by or with the aid of advisers. This led to the exclusion of the minor hosts from regions in which they transmitted parasites to major crops.

Transmission from Early to Late Crops. Most parasites that develop locally, and not by massive influx of wind-borne spores or pests, require several multiplication cycles or generations to reach numbers that can cause economic losses. If suitable hosts are available early in the season, the potential number of such cycles and the danger of losses increases. This raises the question of the disadvantages of growing early crops, where these are not too important but can raise the incidence of parasites on main or late-season crops. The crop protection adviser is in a position to determine the effect earlier crops may have on those sown later, and to direct the attention of growers to this potentially significant communal problem.

2.4.7.3 Bees Have to Be Protected from Pesticide Damage

Another aspect of regional crop protection work is the protection of bees from pesticides applied to a crop or drifting into it from other crops in the vicinity. This requires constant contact with advisers in beekeeping and cropping, with the community of beekeepers, and with growers that require bees to pollinate their crops.

This involves the following aspects (Golan, unpubl.):

a) Up-to-date knowledge of the toxicity to bees of the active ingredients of pesticides as well as of specific formulations. The degree of toxicity to bees should be stated on the label of the container in which the pesticide is marketed, but unfortunately this is not always the case. The adviser should certainly be aware of the general level of toxicity to bees of each important group of pesticides, e.g., he should know that organo-phosphorous materials and insecticidal carbamates are highly toxic, while those based on endosulfan endanger bees much less.

b) Methods of spray and dust application that minimize risk to bees:
- Direct coverage of beehives must be avoided at all cost, and the prominent marking of beehives helps to do so.
- Beehives should be placed upwind of the field or orchard to be treated.
- Water to which bees have frequent access must not be contaminated, e.g. excess pesticide left over in the tank should not be dumped at the heads of irrigation lines or near taps, where puddles often form.

c) Timing of applications: Crops that attract bees should not be sprayed when in flower, but if this is unavoidable, applications should be made while bees are in their hives, i.e. in the evening or at night, before sunrise. If this is not practicable, and crops must absolutely be sprayed in day-time, the hives should be closed, but only for a few hours and with sufficient aeration and supply of drinking water.

d) Cooperation with beekeepers is essential. They tend to consider all pesticide application as a hostile act, but must be convinced that enlightened cooperation will minimize damage. Presentation of suitable, objective information on pesticides, modes of application and precautionary measures to beekeepers therefore falls within the range of tasks of the crop protection adviser.

It is most advisable that beekeepers should be given advance notice of pesticide application in the vicinity of their beehives, especially where applications are made by aircraft. In fact, in some states of the United States, users of methyl parathion are required by law to alert registered beekeepers within a distance of 2 miles of application sites.

2.4.7.4 Regional Pest and Disease Control

Many aspects of control far transcend the boundaries of individual farms and are to be considered regional. Some of these aspects will be discussed here.

Integrated Control. Integrated control is at least in part a regional undertaking. It requires planned timing of the release of hyperparasites and predators, planning of crop areas, e.g. to prevent pesticide drift from one crop from affecting beneficial insects or mites in another, and evaluating the survival and efficacy of such biological control agents in keeping pests in check.

Detection and Avoidance of Parasite Resistance to Pesticides. It has rightly been pointed out by Carlson (1977), that a genetic pool of non-resistant pests or pathogens constitutes a valuable regional asset which the entire farming community must cooperate to safeguard; but with the ever widening use of systemic pesticides, development of resistance of pests and pathogenic fungi has increased sharply. One of the adviser's urgent tasks is to instruct farmers how to retard development of such resistance by sparing use of pesticides against which parasites tend to acquire resistance, and to detect foci of resistance promptly, in time to forestall failures in pest and disease control in the whole region.

Communal Pest Control Operations. The adviser's initiative is often required to get the farming community to organize regional pest control operations, especially against pests that move about freely, such as rodents, fruit flies, etc. This involves establishment of a professional basis for such operations, to aid in the choice of pesticide, the timing of the operation, and the assessment of its short- and long-term effect. But the adviser should not, in general, bear the operational responsibility for regional pest control schemes.

References

Carlson GA (1977) Long run productivity of pesticides. Am J Agric Econ 59:543–548
Hennebery TJ, George BW (1969) Economic entomology – a challenge to our ingenuity. Proc North Central Branch Entomol Soc Am 24:74–81
Palti J (1981) Cultural practices and infectious crop diseases. Springer, Berlin Heidelberg New York
Rondomanski W (1969) Methode zur Prognose der Intensität der Primärinfektion durch den falschen Mehltau der Zwiebel in Polen. Arch Pflanzenschutz 5:89–96

2.5 Pesticide Application: Technique and Efficiency

H. FRANKEL

In modern agricultural production, the application of pesticides is one of the essential components of crop protection.

Most pesticides are extremely potent and have to be applied at minute dosages, at rates of 1000 g or less of active ingredient per hectare. The application of such small quantities poses a number of problems not only with regard to the techniques and equipment employed, but also in the formulation of the active ingredient. Optimal application methods can be defined only when account is taken of all factors influencing the dispersal of the pesticide over its target surface, and its deposition and adherence to that surface. Also to be taken into consideration are the characteristics of the pest, its mobility and site on the host, as well as environmental conditions, such as temperature, relative humidity and air movement.

Table 1 shows those parameters in pesticide application which are amenable to change by the application, and those which are not.

The crop protection adviser has to be familiar with all these aspects, in order to choose suitable techniques of application for a given problem. The choice of pesticide and application techniques is frequently governed by the equipment available.

2.5.1 The Application Target: Crop and Pest

2.5.1.1 The Host Surface Area, Its Shape and Physicochemical Properties

The physical dimensions and surface properties of the application target, which in the majority of cases is the host plant, are of major importance in the processes of deposition and retention. The uniformity of target surfaces in even one crop varies not only in size and shape of leaves, stems and fruits, but also in the physicochemical properties of these parts. Leaves may have a high degree of waxiness on both their surface (cabbage, onion), but may also be smooth, rough or hairy to various degrees on one, but not on the other surface. Moreover, the shape and dimension of even the same plant part may differ significantly (Fig. 1).

A flying insect represents a much larger target than a stationary one, and the wing movement may cause increased deposition.

The shape and physicochemical properties of the crop parts treated affect the deposition and retention of the pesticide, as well as its spread after deposition, and we shall consider these three aspects in turn.

Deposition. Spray droplets emitted from a nozzle undergo a process called droplet transport, which will be more fully discussed in Chapter 2.5.3 of this chapter. On the way to their target the droplets are influenced by a large number of variables and these will determine whether they are collected by flying insects or plant surfaces, by sedimentation or impaction, the latter having special importance for droplets smaller than 50 μm. The collection efficiency of an obstacle in an airstream is defined as the ratio of the number of particles actually striking the obstacle to the number that would strike it, if the air was not deflected. Figure 2 shows the streamlines of flow around a fibre (Dimmick 1969).

In general, collection efficiency increases with the size of the drop and its velocity relative to the target obstacle, and decreases as the obstacle increases in

Table 1. Parameters in pesticide application

Controllable factors	Factors that cannot be controlled
Equipment	Environmental conditions
Technique of application	Transport of droplets or particles after release from nozzle
Pesticide formulation	Collection efficiency of droplets or particles
Timing of application	Pest behaviour
Adaptable crop conditions (e.g. sowing density, irrigation, variety)	Crop characteristics (e.g. upright or creeping habit; leaf area/volume)

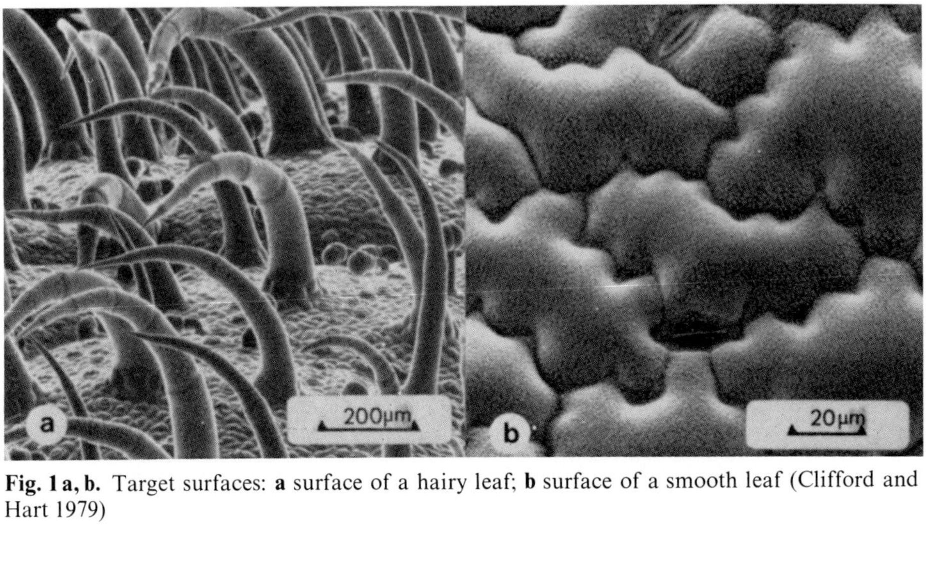

Fig. 1 a, b. Target surfaces: **a** surface of a hairy leaf; **b** surface of a smooth leaf (Clifford and Hart 1979)

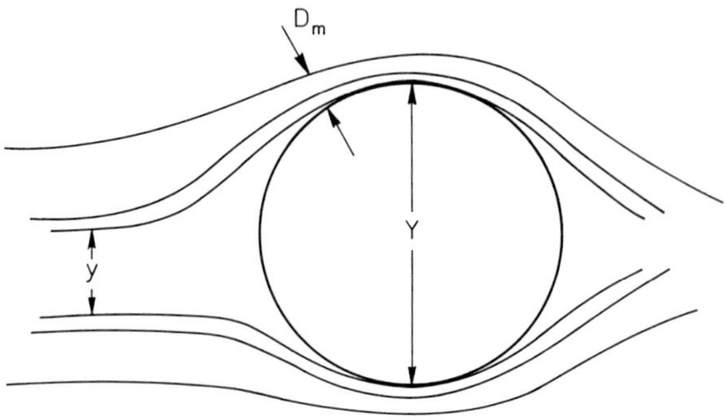

Fig. 2. Streamlines of flow around a fibre. *Y* fibre diameter; *y* effective fibre diameter; D_m stopping distance of a particle

size. Thus, in Fig. 3, at a wind velocity of 3 m s^{-1}, 60% of droplets sized 50 μm will deposit on an object 15 mm in diameter, but 90% of droplets on an object only 3 mm in diameter (May and Clifford 1967).

Solid particles, such as dusts or granules, whilst irregularly shaped, will conform to the same physical laws as drops, but their settling or impaction velocity is strongly influenced by their aerodynamic diameter (da), which refers to the equivalent behaviour of a spherical particle of unit density.

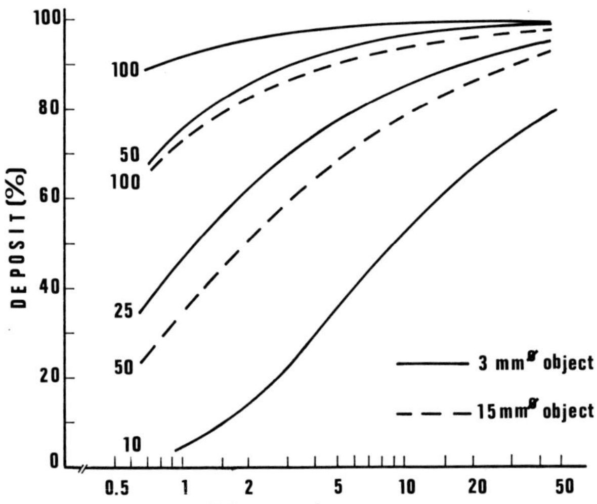

Fig. 3. The theoretical percentage of deposit on two strips of differing diameter (3 and 15 mm) from droplets of four sizes (10, 25, 50 and 100 μm diameter) and velocities ranging from 0.5 to 50 m s^{-1}. (After May and Clifford 1967)

Retention of Pesticide. Spray drops which have reached their target may not be retained, due to run-off or bouncing. Shattering of impacted droplets has been associated with surface roughness for droplets larger than 150 μm. Figure 4 shows a schematic process of droplet deposition and retention. In the wetting of leaves, the angle at which the drop impacts is also important. Ideally, the advancing contact angle (Φ) should be as wide as possible, and the receding angle as narrow as possible (Fig. 4) (Johnstone 1973).

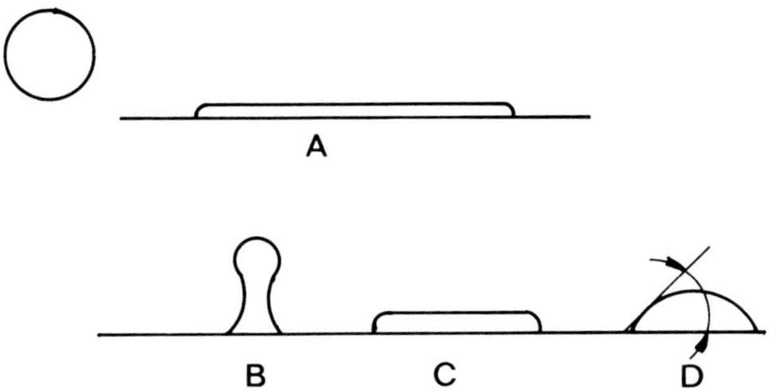

Fig. 4. Schematic representation of the process of droplet deposition and retention. *A* initial spread; *B* retraction; *C* secondary spread; *D* final equilibrium, with indication of the receding angle Φ. (After Johnstone 1973)

It is self-evident that retention to hairy surfaces is greater than to smooth ones. This is especially pronounced where wettable powder formulations or quick-breaking emulsions are employed, as the water diluent will run off the hairy surface, and the solid particles or oil globules will be retained. When fungicides formulated as wettable or flowable powders are applied by dipping in tanks, retention ratios as high as 1:11 have been found (see Chap. 2.5.4).

Dust particles, which normally come in aggregates or conglomerates, may shatter on impact. Coarse dusts deposit readily, but have inferior retention properties; conversely, finer dust particles deposit less easily, but have excellent retention quality. Retention of granules on leaf surfaces is generally poor: large granules will bounce off on impact, while smaller ones may be retained for a given time, until foliage movement causes them to roll off, to be partially collected in leaf junctions, as in maize, or in the "heart" of crops such as beet or cabbage.

Spread After Deposition. As shown in Fig. 4, impacted droplets spread on the target surface. The spread factor (which is the ratio of the diameter of the residue to the droplet's diameter while air-borne) is significantly influenced by both properties of the target surface, and the viscosity and surface-tension properties of the pesticide applied. It can be seen from Fig. 5 that 200 µm droplets with a spread factor of 2, when applied at the rate of 10 litres/hectare will have a final contact area of 1% only, while 50 µm droplets with a spread factor of 4 will have a contact area of 12%.

Most pesticides are formulated by their manufacturers so as to obtain optimal surface tension. The addition of surface-active adjuvants is not advisable except on expert advice and for special purposes.

2.5.1.2 The Position and Orientation of the Host Organ to Be Protected

Topical applications, directed towards specific plant sites, are rare in crop protection. Most application methods (sprays, dust, gas) are non-selective and con-

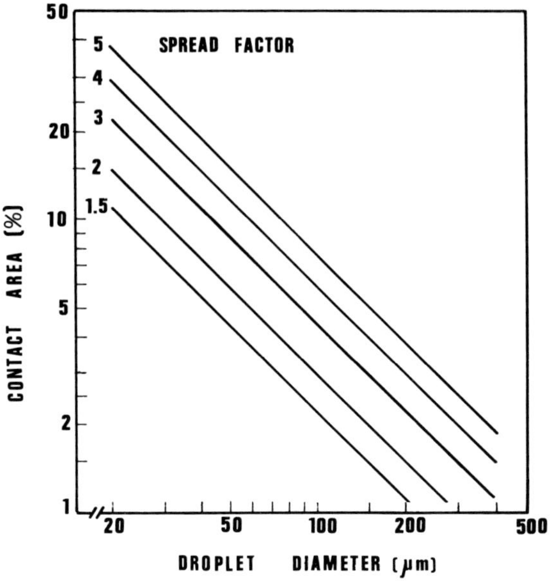

Fig. 5. The theoretical contact area of droplets of various diameters and spread factors, after they have obtained final equilibrium, when applied at 10 l ha^{-1} to a flat surface

taminate to a greater or lesser degree all of the crop and its environment. Thus, in orchards, although the actual target may be the fruit, all other above-ground parts of the tree will by necessity be covered by the pesticide applied.

Certain techniques can be employed to achieve some selectivity in application. Thus to control the European Corn Borer (*Ostrinia nubilalis*), which primarily attacks at leaf junctions, granules may be used that will, in part, roll into such junctions and collect there.

In dense crops, such as well-developed cotton and fruit trees (space crops), certain plant areas are "shaded", and this may present serious application problems. These can frequently be overcome by the use of air-assisted spraying or by high-volume, run-off spraying.

The orientation of the host, whether primarily horizontal or vertical, and its shape, have obvious significance for the deposition process of pesticides. Vertical targets tend to collect more spray or dust than horizontal ones, primarily on their up-wind side (Anonymous 1980). This is due to the tangential fall of droplets or particles, as determined by their velocity and prevailing wind speed (see Chap. 2.5.3). This effect of tangential fall is not marked on horizontal targets such as leaves on vines creeping horizontally, but very pronounced on vertical targets, such as cereals or onions.

2.5.1.3 The Leaf Area Index (LAI) and the Leaf Area Volume Index (LAVI)

The surface area of a crop is defined by its leaf area index (LAI) (Matthews 1975), which represents the ratio of the total area of above-ground parts (leaves, stems, flowers, fruits) divided by the ground area on which the crop grows. A vegetable crop, such as onions, may have a LAI value as low as 0.6 or less, while fruit trees often have indices of 5–6, and a well-developed cotton crop can have a LAI value as high as 7. In actual fact, the LAI may be much larger, as few leaf surfaces are uniformly smooth, and have undulations that may be many microns high or deep. On hairy surfaces, hairs are several microns thick, and sometimes thousands of microns high. It should be noted that LAI values may change greatly as the crop grows, a factor to be duly considered when determining doses of pesticides to be applied to a given crop area.

In crops occupying a large volume of space (space crops), such as fruit trees, bushes, or fully developed field crops, the leaf area/volume index (LAVI) should be considered. This value defines differences in foliage density ratios for a specific plant area, and changes during the growing season. A good example for such changes in field crops is cotton. In evergreen tree crops, such as citrus, the LAI value often stays almost constant once the tree is fully developed; however, the LAVI differs greatly in various parts of the tree and can easily be ten times higher in the periphery of a full-grown orange tree than in the centre of its canopy.

The LAVI value is calculated according to the following formula:

$$\mathrm{LAVI} = \frac{dA}{dV} = \frac{cm^2}{cm^3} = \frac{1}{cm},$$

where A represents the total area of leaf surface, measured in cube shaped volume V.

For representative LAVI values in tall trees, the sample volume dV should not exceed 125,000 cm^3.

For example, let us compare three sections of a citrus tree with a leaf area index (LAI) value of 5. Section I is situated in the periphery of the tree, section II somewhere half-way between periphery and centre, and section III in the centre of the tree. Each sample has a volume (dV) of 50 cm × 50 cm × 50 cm (total 125,000 cm^3). The total leaf area measurement (dA) from sections I, II and III are assumed to equal 25,000, 8000 and 2800 cm^2, respectively. The LAVI values are therefore 0.2 for section I, 0.054 for section II, and 0.0224 for section III. This means that the density of surface area is 3.7 times greater in the periphery than in the area between periphery and centre, and almost 9 times greater than in the centre.

LAVI values greater than 0.1 denote very dense foliage with strong leaf-shading effects, and require special consideration in choosing application techniques and pesticide dosages.

2.5.2 Pesticides and Application

Pesticides may be classified by the way they act on pests and pathogens, as contact/stomach, volatile, penetrating or systemic poisons. It is self-evident that successful control depends on the parasite coming into contact with a pesticide dose sufficiently large to inactivate or kill it before it causes economic damage. The processes involved, specifically the mode of action of pesticides, will be briefly discussed.

Stomach Poisons. The efficacy of stomach poisons applied to plants is related to the amount of residue retained on a treated surface, and to the size of that surface ingested by the pest, with the host tissue acting as a bait. Any factor that affects the feeding behaviour of the pest influences the efficacy of the treatment. Among these factors are:

a) Temperature: excessively high or low temperatures that may tend to reduce the rate of feeding.
b) Stage of pest development: (1) adult stages of pests such as beetles may consume relatively large amounts of tissue, including upper and lower leaf surfaces; (2) the feeding behaviour of the larvae, grubs and maggots changes as they pass through successive instars: when young, they often feed only on the under-surface of leaves and these frequently carry less spray deposit than the upper surface. In later instars they consume progressively more tissue of both leaf surfaces, until they approach pupation, when feeding declines. At this stage larvae of certain moths, e.g. *Spodoptera littoralis*, also become less sensitive, weight for weight, to some organo-phosphorus poisons.
c) Repellency: some stomach poisons repel pests to some extent. This obviously limits their usefulness, and can be overcome when they are mixed with highly attractive material and used as baits (see Chap. 2.5.4).

Contact Poisons. Where fungal or bacterial pathogens are concerned, efficacy of treatment with contact chemicals, applied as protectants, has long been held to be directly related to the degree of coverage attained, especially at the site of penetration, which is often on the underside of leaves or on other plant parts not easily accessible. Cover requirements were thought to be a thin continuous film over the entire leaf surface. Fortunately, this is not true, as it cannot be achieved. Most protectant fungicides undergo either an external translocation or redistribution process, as in the case of copper spray applied against potato blight (*Phytophthora infestans*), or have an area of influence around the discrete particle deposit.

It may be generalized that, for the control of pathogens, the spray cover aimed at should be denser (more droplets per cm^2) than when the aim is control of highly mobile insect pasts. But proof that fungicides can successfully control diseases while covering only a small, non-continuous part of the treated foliage, is to be found in the success of reduced volume applications, based on deposition of discrete droplets, in controlling most diseases.

Contact action on insects and mites depends to a large extent on the pest's behaviour, especially its mobility and "pick-up" characteristics. A mobile pest, in its movement over the host surface, evidently stands a better chance of coming into contact with more spray deposit than a stationary or sessile pest, especially when the latter is located on the leaf's underside. This is particularly true where coverage is not uniform. The "pick-up" of deposit by a pest on a crop sprayed with contact insecticides is chiefly related to the proportion of the pest's tissue that comes into actual contact with the deposit. Thus, a larva creeping over a sprayed surface will pick up on its underside far more deposit than a legged insect of equal size, which only has the ends of its legs come into contact with the deposit (unless the insect "cleans" its legs by passing them through its mouth parts). It should be noted in this context that almost all contact poisons are even more lethal when ingested. To give an example of the relation of pest behaviour to efficacy of coverage, it may be said that the cover density adequate for control may be only 3–4 droplets per cm^2 for some larvae, but as much as 100 droplets cm^{-2} for sucking insects, or even 300 droplets cm^{-2} for sessile armoured scales.

Volatile Pesticides. Volatile pesticides can be described as contact poisons which, due to the small size of their molecules and high vapour pressure, act in a gaseous state. Because of their small molecules they have high mammalian toxicity, and must be used with proper precautions. Highly volatile pesticides, such as fumigants, are used in buildings, storage facilities or greenhouses, primarily because they have no, or only very short, residual effects and because of their good penetration properties. Some contact poisons used in crop protection act, at least in part, by virtue of their volatility, which increases their sphere of action beyond the plant surface actually covered. Volatile sprays may act on moths and other flying insects not only when they alight on the plant, but also in flight. Thus, it has been found in Israel that aerial application of a volatile insecticide at midday with high temperature conditions resulted in better field control of the adult tobacco white fly (*Bemisia tabaci*) on cotton than treatments applied at dawn, although the pesticide residue recovered from dawn application was higher by about 30%. In this case, control, at least in part, was probably achieved by killing the insects in flight.

One of the criteria for pesticides used to control pests and pathogens in the soil is their efficacy as fumigants. Successful treatment, however, depends, not only on their specific weight and vapour pressure, but also on the porosity and organic matter content of the soil, and on proper preparation of the soil for treatment. It should be noted that fumigant treatment of soil may sometimes leave residues which, unless washed out, may be harmful to the succeeding crop. Handling volatile pesticides always involves some risk to the operator, and these materials should be handled only by personnel trained to observe safety precautions.

Penetrating Pesticides. These are stomach or contact poisons which partially penetrate the cuticle into underlying tissues ("Tiefenwirkung"), but are not translocated to other parts of the plant. Pesticide and formulation, as well as plant surface properties, determine penetration rate. The residues of such pesticides are partially protected from weathering. They may also act on pests feeding on plant parts difficult to cover, e.g., the underside of leaves, as the pest reaches the tissues penetrated by the pesticide.

Systemic Pesticides. These are mostly contact poisons, and in the case of insecticides also stomach poisons, that are taken up by the host tissue to which they have been applied, and are then in part translocated upwards and sideways (but generally not downwards) in the plant. With these pesticides, uniformity of coverage is obviously less important. When applied to the root or stem, the material will therefore be translocated to upper plant parts; however, while the leaf deposit will translocate throughout the leaf, note that it will not descend to the petiole and hence cannot rise to upper tiers of leaves. It is therefore important that coverage be dense enough to deposit sufficient material on each plant part that needs protection. Factors of prime importance in determining efficiency of systemic pesticides are their rate of penetration and translocation, as well as their rate of retention and dilution within the plant system. Systemic pesticides differ widely in these properties; their mode and rate of application must therefore be determined specifically for each pest and crop, since surface properties of the latter may greatly affect the uptake process.

2.5.3 The Movement of Pesticide Particles, as Influenced by the Environment, and the Danger of Drift

Most pesticides are dispersed as clouds of small droplets or particles, the behaviour of which after their release into the atmosphere is strongly influenced by ambient conditions. Unless placed into an air-carrier stream, the primary distance of travel of such particles, due to the emitting forces, is short and their movement is governed principally by gravitational fall. Table 2 (see p. 142) indicates the terminal velocity and stopping distance of water droplets at 20 °C (spheres with a unit density of 1) (Davies 1966). The time required by a droplet of known volume or diameter to reach the target when released from a given height is therefore known. Air movement, such as wind or turbulence, will affect the flight path of the particle, causing it to fall at a tangent. The distance it travels will be the product of prevailing wind velocity, height of release, and time of fall (Fig. 6). Droplet evaporation and air turbulence will obviously increase this distance.

It is also clear that under windy conditions, or when released from great height, small particles or droplets are drift-prone and may be carried not only outside the treatment swath and target area, but travel many hundred metres or more.

The time required for a water droplet to evaporate to extinction is a function of ambient temperature and of the moisture content of the air (relative humidity). The speed at which water drops evaporate increases markedly with decrease in relative humidity and rise in temperature. The theoretical life-time of such a droplet may be calculated by Amsden's equation (Amsden 1962):

$$t = \frac{d^2}{80_{\Delta} T},$$

where t is the life-time (in s), d is the diameter of the droplet (μm), and T is the wet-bulb depression (°C).

In most run-off spray applications, the percentage of formulated pesticide is as low as 0.1 – 0.3%, and the evaporation potential of droplets containing so low an amount of pesticide does not significantly differ from that of water droplets.

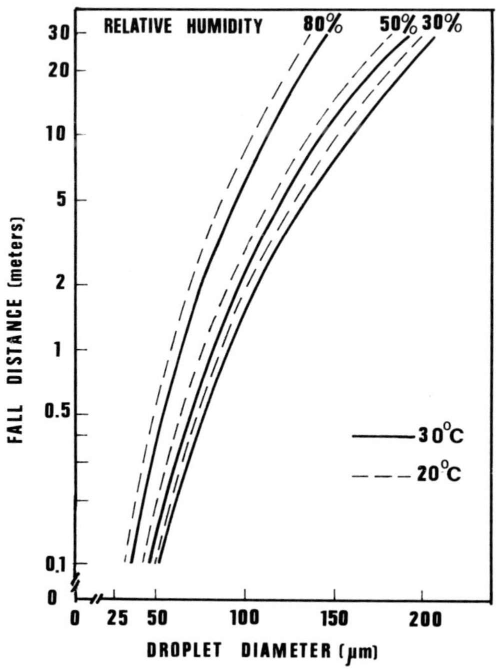

Fig. 6. Theoretical fall distances (until extinction) of water droplets with various diameters, at two levels of temperature and three levels of relative humidity

The theoretical fall distances of water droplets until their extinction is shown in Fig. 6 in relation to three different ambient conditions (Anonymous 1980).

All reduced volume applications (discrete droplet sprays) with the exception of ULV, are made with water-based sprays. The percentage of formulated pesticide used in these sprays may be as high as 10–30% of the total volume. At such concentrations, the evaporation potential of the spray drops may increase, and evaporation rates may be faster than those of water drops. Similar considerations apply when wettable powders are used at high concentrations, as a considerable mass of the droplet then consists of solid matter.

This evaporation potential may have an important practical consequence, especially when sprays are applied under hot and dry conditions: water-based sprays with small droplets (volume median diameter 150–175 μm or less) should not be applied when the wet bulb depression is greater than 5 °C and temperature is equal to or above 30 °C. If spray application begins early in the morning, under cooler and more humid conditions, their application should cease when atmospheric humidity decreases beyond the above value and temperature rises. Increasing drop size to a volume median diameter of 200–250 μm halves the number of droplets applied, but will permit spraying up to a wet bulb depression of 8 °C, at 35 °C (Johnstone and Johnstone 1977). Values of wet bulb depression are determined with the aid of a psychrometer, a dry and a wet bulb thermometer, which should be available to every adviser concerned with spray efficacy.

The effect of air movement, caused by winds or turbulence, on the deposition of droplets, i.e. their collection efficiency, has been referred to in Chapter 2.5.1 and Fig. 3; collection efficiency has there been stated to increase with particle or droplet size and their velocity relative to the target obstacle, and to decrease as the obstacle increases in size.

Some of the processes described above as influencing the behaviour of spray droplets may not be familiar to crop protection advisers and are indeed complex in their interaction. An attempt has therefore been made to present the processes involved in a schematic diagram (Fig. 7), in which the droplet is followed from its production and release until it is deposited and retained on the target. The factors affecting each stage are indicated, and the figure shows that some stages are influenced by a large number of parameters.

The disintegration of a liquid by the processes of atomization may be called droplet production. Liquid emitted from an atomizer, such as a spray nozzle, takes the form of a sheet or of ligaments or drops. The physicochemical properties of the sprayed liquid affects the atomization process and the size characteristics of the spray (Fraser 1958). Both sheets and ligaments break very rapidly into drops which possess velocity and direction. The distance a drop travels, due to efflux velocity in still air, is termed the stopping distance or initial range, and is defined as the displacement over which the horizontal component of droplet velocity falls to zero (Hinds 1982). The stopping distance (Table 2) gives some indication of the range over which efflux velocities can have an overriding influence on the width of the band sprayed and on the travel of spray drops as they leave the nozzle. This is of practical importance for coarse or medium sprays, with droplets sized 300–500 microns, so that such droplets can be directed onto the target from a short range (placement spraying). Efflux velocity, however, has a proportionally negligible effect on travel and deposition of small droplet sprays, or when the sprays are air-carried (Anonymous 1980).

Table 2. Terminal settling velocity (V_{TS}) and stopping distances (d_s) for projection velocity of 1 m s^{-1}, for droplets of diameter (d_p) and sp. gr. 1, in the size range 1–1000 μm

d_p (μm)	Wt (μg)	V_{TS} (cm s^{-1})	d_s (cm)
1000	524	386	39.2
500	65.5	202	20.4
300	14.1	117	11.7
200	4.2	70.8	7.14
150	1.7	47.2	4.69
100	0.5	24.8	2.55
50	0.065	7.53	0.734
30	0.014	2.72	0.275
20	0.004	1.21	0.122
10	0.0005	0.305	0.0306
5	0.00007	0.0776	0.00795
1	0.0000005	0.0035	0.00036

Fig. 7. Interacting parameters which influence a droplet's flight path from the spray nozzle to it's retention on the target

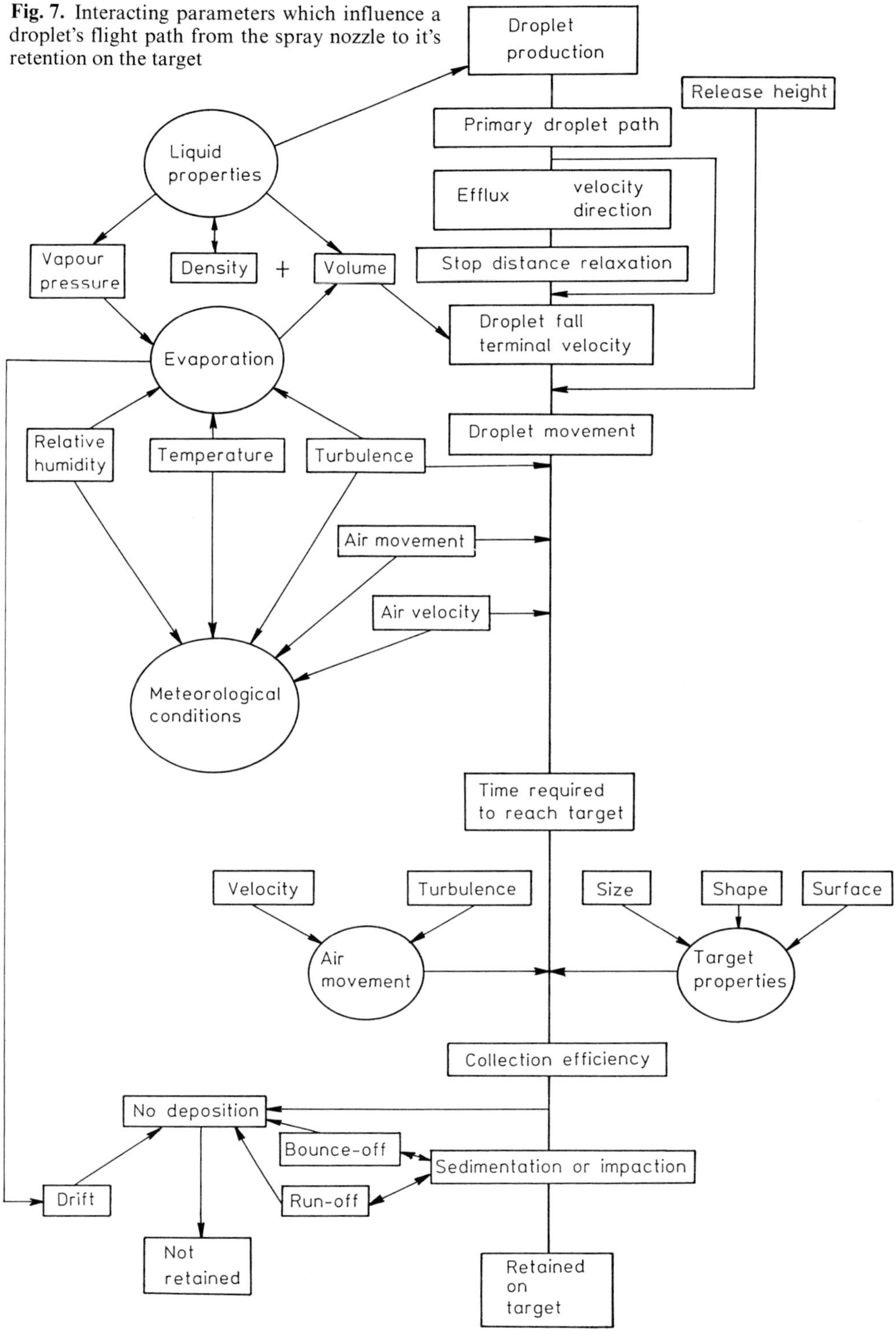

Continuing its flight process, as depicted in Fig. 7, the droplet or particle then begins to adjust, or "relax" its velocity to new forces, including gravitation; the time it requires for this adjustment is termed "relaxation time" (Hinds 1982).

The rate at which a small droplet falls is determined by the balance of gravitational and viscous forces in such a way that a steady fall velocity or terminal velocity is rapidly attained. The terminal velocity of water drops in the 1 – 1000 μm range, and their stopping distance are presented in Table 2.

As already stated, the droplet's path is strongly influenced by air movement, and this influence is inversely proportional to its inertia. Under climatic conditions conducive to evaporation, the terminal velocity of the droplet will decrease continuously as a function of its diminishing volume, and hence the time it takes to reach the target, if at all, will increase considerably; side winds will carry these droplets outside the target area (drift). When they eventually reach the target area, some of the meteorological conditions which have previously affected the droplet's flight path, will influence the deposition process, i.e. sedimentation or impaction. The relative importance of these two deposition processes may be estimated by the ratio of terminal droplet velocity/wind velocity, which defines the impaction or collection efficiency of a target in an air stream laden with small droplets (see Fig. 3) (Anonymous 1980).

Deposition on the target does not necessarily imply retention. In the use of techniques which cause complete wetting of the target surface, large volumes of liquid will run off, while coarse droplets may bounce or roll off, as has previously been explained. Eventual retention of the droplet on the target is affected by three crop parameters (size, shape and surface properties) and by meteorological factors such as air velocity (wind) and turbulence.

2.5.4 Pesticide Distribution Techniques

To visualize the problem of application involved in distributing minute quantities of chemicals, let us consider the case of applying 1000 gr (10^9 μg) of pesticide to a crop standing on 10,000 m^2 of ground with an LAI of 5. Even application would result in a deposit value of 2 μg/cm^2; if uniform and having a density of one, the thickness of the deposit layer would be 0.02 μm or 2 ten-thousandth of a millimetre (Green et al. 1977). This is slightly more than a monomolecular layer and manifestly impossible to achieve even for liquid films, let alone for wettable powders or dusts. For the reason discussed in Chapter 2.5.5, continuous coverage of this sort is also unnecessary.

2.5.4.1 Spray Application Techniques

Spraying may be defined as the dispersal of a liquid after its break-up into individual droplets, by an atomizing device such as a nozzle.

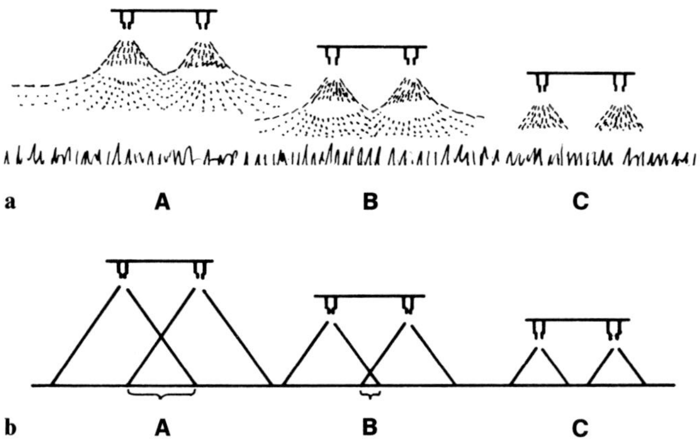

Fig. 8a, b. The proper height of nozzles for boom application by spray cloud and placement spraying techniques: *A* too high, small droplets may drift beyond the target, and in placement spraying overdosage is applied in the centre; *B* correct height; *C* too low, therefore, no overlap, but overdosage in the bands covered in placement spraying

Spray Clouds. Most agricultural atomizing devices form spray clouds. The sizes of droplets in the cloud depend on nozzle type, liquid pressure, and physicochemical properties of the spray liquid. Droplets are released above the target and impinge on it due to impaction or sedimentation. This technique is widely employed in spraying with boom sprayers and from aircraft. Nozzles of the hollow cone type are generally used for the application of fungicides and insecticides. They should be placed at a uniform distance (40–45 cm) on the spray boom, and their height above the crop should then be about 60 cm, to permit the formation of a uniform cloud of droplets (Fig. 8). In such application, the effect of the speed at which the sprayer travels is mostly negligible as far as the droplet's deposition efficiency is concerned.

For special problems, e.g. improving coverage of tall field crops (cotton, maize), nozzles are affixed to vertical booms descending from the horizontal boom. Since droplets are then emitted from nozzles passing at right angles to the plants, the droplets have to be sufficiently large to travel the required distance to the target. In this case the forward speed of the tractor is relevant, and should not be too high.

Placement Spraying. Spray drops can be impelled towards the target by imparting kinetic energy sufficiently large to carry them over the distance required to make them impinge on the target. The distance over which a droplet of a given size can be thrown is known as its "stopping distance" (Hinds 1982) (see Table 2). Even at high initial velocity (200 m s^{-1}), the stopping distance for small droplets (up to 100 μm) is relatively short and can be measured in centimetres, while large droplets have longer stopping distances when impelled at comparable velocities. Larger droplet sizes (250–400 μm) with very high efflux velocities, are therefore used in application by spray guns to treat a wide swath or a tall tree, when large throw distances are required.

Spraying of soil surfaces or of very low vegetation is carried out with spray booms equipped with fan-type nozzles; these produce a sheet-like spray, where most of the liquid volume is applied to the centre of the sheet and decreases towards the edges. A proper overlap between nozzles, amounting to approximately 25% on either side, should therefore be achieved to ensure uniform volumetric distribution. The factors on which such overlap depends are (a) the angle at which the spray is emitted from the nozzle, as stated by the manufacturer in relation to the liquid pressure; (b) the distance between nozzles and the boom; and (c) the height of the nozzle above the surface sprayed (Figs. 8 and 9). It may be noted that fan nozzles with uniform volumetric distribution, and others with a minimum of spray emitted at the centre, are available but are used only for specific purposes. Thus E nozzles for even spray distribution are employed primarily for band spraying, e.g. application while the crop is being sown (Matthews 1975), most commonly in the application of herbicides.

Controlled Droplet Application (CDA). In CDA spraying the atomizing device produces a somewhat narrower spectrum of droplet sizes, as compared to that produced by conventional hydraulic nozzles. Spinning cups or discs are most frequently used for this purpose, and are employed for reduced volume application (5–30 l ha^{-1}). The primary advantage of CDA is the reduction in the amount of small, drift-prone droplets at very low feed rates, when the required volume median diameter of droplets is in the 200–250 μm range, and the nozzle orifice is relatively large. Such applications are mostly made in herbicide treatments. When made for the control of pests and diseases, where droplet sizes in the range of 80–120 μm are used, the droplet spectrum produced by the spinning cups is not narrower than those produced by conventional nozzles. But they have the advantage that the swath covered by a single nozzle is about 1.20 m wide, equalling the swath of two to three conventional nozzles; this permits the use of nozzles with relatively large orifices working at low pressure, and hence makes the output of small volumes possible.

Electrostatic Spraying. This relatively new application technique, still in its experimental stage, is based on the production of spray clouds in which individual droplets carry a positive charge. They are attracted to and deposited on

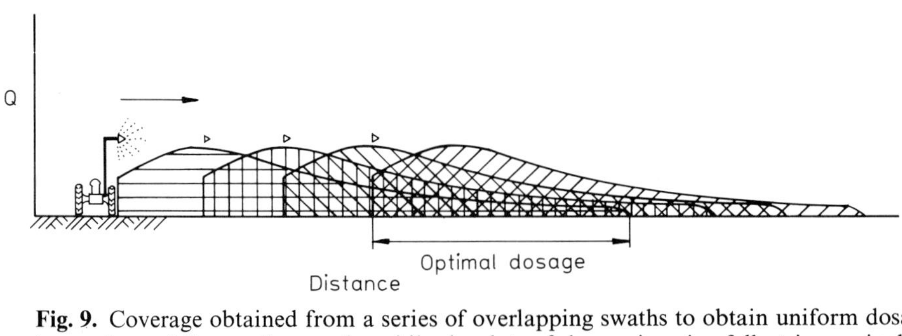

Fig. 9. Coverage obtained from a series of overlapping swaths to obtain uniform dosage. The larger drops fall near the nozzle, while droplets of decreasing size fall at increasingly longer distances. (Q = quantitative pesticide deposit per unit area)

plant surfaces which have a negative charge, if and when the attraction forces are greater than the momentum of the droplets' kinetic fall. The advantages of this technique remain to be proved, especially where application to dense foliage is concerned.

Air-Assisted Spraying. Spraying techniques in which the droplets emitted are carried towards their target by an airstream are termed air-assisted spraying. These techniques may be classed into air-carried placement spraying and drift spraying (including fogging).

Most agricultural air-assisted sprayers employ either radial or axial fans, which differ in their performance: radial fans deliver a relatively small volume of air at high velocity, while axial fans supply large volumes of air at low pressures.

In *air-carried placement spraying*, the carrier air stream has a three-fold task: (a) to carry the droplets to the target; (b) to turn the foliage and ensure adequate penetration into the plant canopy; and (c) to assist in droplet deposition.

When air is discharged into the atmosphere, it rapidly loses velocity with distance from the outlet, the rate of loss depending on the shape and area of the outlet and on the airstream's initial velocity. Thus it has been shown that the initial velocity, at the nozzle, decreases to 40% at a distance 20 times that of the nozzle's diameter, and to only 10% at a distance 90 times that diameter, when sprays are applied by a radial fan, such as used in knapsack sprayers. If, therefore, the initial velocity is 50 m s^{-1} at an outlet with a 5-cm diameter, then at a distance of 100 cm the velocity will decrease to 20 m s^{-1} (Matthews 1975).

Formulae have been developed to indicate the air velocities obtaining at various distances from *stationary* outlets; however, these do not apply to moving air sprayers, where travelling speed and differing horizontal, as well as vertical distances between outlet and target, have to be taken into consideration, and where the leaf area/volume index (LAVI) also affects the velocity with which the foliage is penetrated by the airstream.

The velocity of a highly turbulent air mass that encounters a dense leaf canopy will be severely reduced by the "shock-absorber" type of behaviour of the foliage. The airstream will also be deflected by the multi-plane surfaces, the flutter of which further increases air turbulence and multi-directional movement. Large air masses, such as those generated by axial blowers, do not penetrate well into dense foliage with a LAVI value of more than 0.1. Slow-moving air mass "closes" the foliage, and produces shade effects, with peripheral leaves preventing central leaves from being covered, in addition to causing run-off from the peripheral surfaces.

As a rule of thumb, an air velocity of 5–6 m s^{-1} (at the target) is considered optimal for deposition of small droplets, such as used in reduced volume application. For application of pesticides to dense trees and bush canopies, high-velocity air sprayers (radial fan sprayers) are preferable. As stated above, in all cases air velocity at the target at a given distance or height is greatly affected by the speed at which the spray equipment travels and, in the case of manually operated outlets (such as knapsack mist-blowers), by the movement of the outlet (see Chap. 2.5.5).

A common error, when operating air-assisted sprayers where air velocity is utilized in the atomization process, is overfeeding, i.e. putting in more liquid than the airstream can efficiently atomize. The weight of one cubic metre of air is approximately 1 kg. Weight/weight ratios of air to liquid should not exceed 5–20 if a low velocity air stream (50–150 m s^{-1}) is used for the atomization of the liquid, as is almost always the case in manually operated knapsack sprayers. Therefore, a sprayer with a volume capacity of 5 m^3 air per minute should not be fed more than 200–1000 ml liquid per minute (Fraser 1958). Since all knapsack sprayers have emitting air speeds of approximately 50 m s^{-1}, liquid feed rates should not exceed 250 ml m^{-1}.

Drift spraying is the release of droplets into a prevailing wind. Movement of the droplets after their release depends on their mass, on wind velocity and direction, on the height of droplet release above the crop, and on air turbulence. This technique is usually employed for insect pest control for agricultural and public health purposes. Specially formulated pesticides with minimal evaporation characteristics are thus applied, usually at ultralow volumes, when residual effects are required.

All agricultural sprays are made up from droplets in a wide range of sizes. Due to their mass, large droplets carried by the air stream fall out more quickly than small droplets, resulting in an uneven volumetric distribution across the spray swath. The effective field swath is therefore built up from a series of overlaps (see Fig. 9) (Matthews 1975). Application by drift spraying may be made by diverse equipment, including aircraft, motorized ground equipment, and even hand-operated sprayers rigged to produce spray clouds with droplet sizes of 40–80 μm, which are released upwind of the target. Use of this technique is limited. Since it is not possible to control drift, pesticides with high mammalian toxicity may not be employed in a populated area, because of the danger to man and beast, or in mixed agriculture with small fields, where non-target crops may be contaminated. Extreme caution must be taken in communities in which aquaculture is practiced, as some species of fish are highly sensitive. The danger of environmental pollution is a prime consideration in drift spraying, and this frequently should override all other considerations. Herbicides should never be applied by drift or airblast spraying, unless it is expressly permitted.

2.5.4.2 Application of Dusts and Granules

The advantage of using dry pesticide formulations, such as dusts and granules, lies in the fact that they do not require further dilution before application. Most dust formulations consist of a relatively small percentage of active ingredient (5–10%) incorporated in an inert carrier by mixture, absorption or adsorption. The production, packaging and transport of such formulations is almost invariably more costly than that of formulations for spraying.

Dust. The use of dust formulations has declined significantly over the past three decades. The reason is to be found not only in the development of sophisticated spray equipment, which permits more accurate metering and placement of pesticides, and of spray formulations with better retention and cover properties, but also in the drift and inhalation hazards inherent in the use of dusts. Moreover, a great number of pesticides is no longer available in dust formulations. Dusting is still used in specific crop/parasite situations, e.g. in sulphuring grapevines: sulphur is often cheaper than other chemicals recommended for powdery mildew control. In addition, dusts are used for treatment of small batches of seeds, bulbs and corms.

The distribution of dusts, normally at the rate of 25 – 30 kg per hectare, is relatively simple. But this method of pesticide application is recommended only where no alternatives are available. Modern dust formulations consist of prime particles smaller than 325 mesh or 44 μm in size. In storage these tend to aggregate and agglomerate. While coarse aggregates deposit readily, they are also easily eroded by wind or foliage movement. A typical picture of a duster moving through the field shows a large dust cloud moving through the foliage. In most cases this cloud consists of large aggregates, which fall out after a short distance, and of prime particles and small satellites. The latter make it appear as if a large swath of foliage is treated successfully, though in actual fact they do not deposit.

Since the physical properties of different dust formulations differ widely, home mixing of two formulations should not be attempted.

Granules. Granules are formulations which consist of small pellets of dry, inert material in which the pesticide has been incorporated. These pellets come in a wide range of sizes. Depending on their specific weight and mesh size, there may be 300,000 granules of 8/15 mesh (1 – 2 mm diameter) per kg, or as many as 800×10^6 granules of 80/120 mesh, the latter in sizes of the order of 0.12 – 0.08 mm, and these are defined as micro-granules. At the smallest mesh size (80/120), the number of pest granules per square metre will be approximately 2600 times greater than at the 8/15 mesh size. Granules were developed in order to reduce drift problems: they should have a very low dust content and application machinery in which the metering device causes grinding should not be used.

The number of pesticides available in granule formulations and their use has increased considerably in recent years. They are used principally to control soil pests as well as aphids, stem borers and maggots, and serve for the formulation of baits. Granules may be applied by knapsack blowers or dusters, and by aircraft. They may also be incorporated into the soil or broadcast by hand and, if soluble, are sometimes applied in irrigation water.

Optimal granule sizes depend on the pest to be controlled and on the equipment to be used for application. In bait application, large granules are applied for three reasons: (a) dense coverage is not required; (b) large granules break down more slowly and may thus remain active for a longer period; and (c) small particles are often covered by dust and are then not available to the pest. In dry climates, such baits are normally applied after irrigation, to make them more palatable.

The use of microgranules is often preferred for the control of insect pests, such as the maize borer, which attack plants at leaf junctions. The collection efficiency for microgranules is three to four times greater than that of large granules. But the use of microgranules is evidently limited to some extent by the availability of suitable application equipment. Microgranules are most efficiently distributed in the airstreams of pneumatic systems with appropriate metering devices. Because of their small mass, these granules should not be distributed by centrifugal spinners, such as used for fertilizer spreading; nor should they be spread from aircraft, as considerable swath displacement may occur at the height (10 m) at which granules are at present spread.

2.5.4.3 Fumigation for the Control of Soil-Borne Pests and Diseases

There are four common methods of applying pesticide to soil: by dissolving them in irrigation water, incorporation of granules in the soil, injection of liquids, or dispersal by gas. In every case, the process of dispersal in the soil is aided by the material's vapour pressure, which effects secondary dispersal to the site of action. The vapour pressure of fumigants makes it necessary to seal off the soil surface after application. This is usually effected by light irrigation, but this practice results in poor action of the fumigant in the upmost millimetres of soil, and is therefore acceptable only where disinfection of the top layer is of relatively little significance. The use of fumigants with high vapour pressure, such as methyl bromide, requires that the soil surface be sealed effectively by a quasi-impermeable layer; for this purpose, plastic sheeting is used.

Application of fumigants to soil is an expert operation, and will not be dealt with in detail here.

2.5.4.4 Dipping, Drench-Sprays and Spray Disinfection

Pesticides are widely applied to vegetative propagation material (set tubers, corms, rhizomes), seed, and fruit to be stored, to control externally borne pathogens and pests.

To obtain results:

a) The material to be dipped should be as free as possible from adhering soil and other impurities.
b) After dipping for the prescribed period, excess liquid must be allowed to drain back into the tank.
c) The strength of the dip must be kept up to the level prescribed, with due consideration of the amount of pesticides that may have been selectively removed by the plant material dipped.

Where emulsifiable or soluble pesticides are used for dipping, the ratio of liquid: pesticide retained on the plant material is constant, i.e. to top-up the bath after several dippings, pesticide is added at slightly above the concentration originally used. However, this does not apply where suspensions of wettable powders are used for dipping, as is often the case in the use of fungicides. Retention of suspended particles on treated surfaces is then selective, and not linear to the rate of liquid withdrawn from the bath. In fact, the ratio of liquid: pesticide withdrawal has been found to be 1:6 for potato tubers dipped in Terrachlor, and as much as 1:11 for citrus fruit dipped in TBZ suspensions. It is therefore of utmost importance that where suspensions of wettable powders are used for dipping, the concentration of the topping-up suspension is adjusted to an appropriate level, often as much as five to ten times that of the original suspension (see Chap. 2.5.1).

Spray disinfection techniques have recently been introduced and replace disinfection by dipping to some extent. The sprays are applied by procedures similar to run-off or to very low volume application. When plant material is to be subjected to total wetting (drench-spraying), it passes through a spray chamber, is then made to rotate on its own axis on a roller belt, and is drenched with the pesticide at a concentration identical to that used in dipping. Excess liquid is permitted to drain back into the main tank, from which it is recirculated. Retention mechanisms and distribution patterns on corms or tubers treated are identical with those after dipping.

Very low volume applications are also made in spray chambers (aerosol chambers), but the pesticide is applied at a rate corresponding to the area of treated surface. One ton of seed potato tubers, which may be considered to have an area of 150 m^2, is treated with about 180 cm^3 of concentrated pesticide, which equals the spraying rate of 12 l ha^{-1} of a field crop. After application at so low a rate, no subsequent drying is required.

A further advantage of this application method is that it permits the use of micronized wettable or flowable powder formulations, since the deposition constant is not affected by the physicochemical properties of the material treated. Kinetic energy nozzles (rotary atomizers), at feed rates of approximately 30 ml min^{-1} and with droplets of 80 μm volume median diameter, are usually employed. The number of times the tuber or corm rotates on its own axis while passing through the chamber is critical, and should be not less than six to seven rotations. When the treatment is applied properly, droplet density should be 350–500 per cm^2. It has been demonstrated that for the control of fungal and bacterial diseases, the results obtained with very low volume spray application of specific chemicals equal or better those obtained by dipping.

2.5.5 Rates of Application and Uniformity of Coverage

2.5.5.1 Rate of Application

The volume of spray applied to a crop largely determines the cost of application. Hence the obvious desire to reduce the volume of application as much as possible. Table 3 summarizes the accepted definitions of volumes per unit crop area (Anonymous 1980).

"High volume" is a term that indicates that the crop is wetted over almost all its surface. The pesticide is therefore applied in a highly diluted form. The quantity of chemical retained on the target surface is determined by its concen-

Table 3. Definition of spray volumes per hectare

	Bushes and trees	Herbaceous crops
High volume (H.V.)	$1000\ l\ ha^{-1}$	$700\ l\ ha^{-1}$
Medium volume (M.V.)	$500-1000\ l\ ha^{-1}$	$200-700\ l\ ha^{-1}$
Low volume (L.V.)	$200-500\ l\ ha^{-1}$	$50-200\ l\ ha^{-1}$
Very low volume (V.L.V.)	$5-200\ l\ ha^{-1}$	$5-50\ l\ ha^{-1}$
Ultra low volume (U.L.V.)	$5\ l\ ha^{-1}$	$5\ l\ ha^{-1}$

tration in the spray liquid and by the interaction of adhesive forces due to the physicochemical properties of the liquid and of the target surface. Since complete wetting of all plant surfaces is aimed at, excess quantities of liquid are applied and drip off the treated surfaces. Thus the amount of liquid retained on 1 ha of citrus trees with a LAI of 4 – 5, is only 200 – 300 l after the crop has been sprayed with an oil emulsion at the rate of 20,000 l ha^{-1}; this equals a retention rate of 1 – 1.5% of the liquid applied. The quantity of liquid necessary to wet all foliage surfaces of most tree crops, not including the coarse drops formed at leaf tips or along the veins, is close to 60 – 80 l per hectare.

It may be concluded that the chemical dose retained on surfaces wetted with solutions or emulsions is not increased by using excessive volumes of diluted chemicals, after run-off has been reached. However, the use of high volumes may cause better coverage of those parts of the host that are not easily accessible in spraying. But when suspensions, such as those of wettable powders, are applied, large volumes exceeding those needed for run-off may achieve slightly higher retention; this is due to the physical properties of the particles, e.g. their size and shape, and the properties of the target, as some additional particles adhere and are retained, while excess water carrier runs off.

Because of the wastage involved in their use, high volume applications are therefore recommended only under very special circumstances, where it is essential to wet plant surfaces as completely as possible in order to obtain control even of pests and pathogens in places accessible only with difficulty.

"Medium volume" is a loosely defined term which usually indicates that the intention is to wet the crop over almost all of its surface. The concentration of chemicals used is still dilute. Medium volume applications are primarily used in conjunction with sprays carried by an air stream. Small droplets, whose cover efficiency per volume/number ratio is greater than that of large droplets, are carried by the air stream to the target which they would not otherwise reach. Medium volume application is therefore an improved method of run-off spraying, and retention of pesticides on the treated surface is identical to that described for run-off spraying.

Reduced volume applications, such as L.V., V.L.V. and U.L.V., are spray techniques in which no attempt is made to wet the whole target surface. The droplets applied are, instead, meant to remain discrete on the target and do not coalesce after deposition. The amount of chemical retained on the target surface is therefore not affected by the properties of that surface to the same degree as in run-off spraying. The principal parameters determining the quantity of chemicals retained will be the concentration sprayed and the target's collection efficiency, for a given period of exposure. The time over which the target is exposed to droplet clouds or to air streams carrying the droplets is of decisive importance in determining the density of deposit and hence its quantity. The rate at which reduced volume sprays, used at relatively high concentrations, are applied must be constant; otherwise overdosing may cause phytotoxic damage to the crop or, conversely, sublethal doses may induce pest resistance.

The number of droplets that can be produced from a given volume of liquid is inversely proportional to the cube of the droplets' diameter. Table 4 indicates the amount of spray liquid required to apply spray at a density of 10 droplets cm^{-2} uniformly to a flat surface (Matthews 1975). It is apparent from the table that the leaf area index (LAI) is an important consideration in deciding on the

Table 4. Spray volume ($l\ ha^{-1}$) required for a cover density of 100 droplets cm^{-2}, for different droplet sizes

Droplet diameter µm	Spray volume $l\ ha^{-1}$
10	0.005
30	0.14
50	0.66
70	1.8
100	5.2
150	17.7
200	41.9
250	81.8
500	654.7

Table 5. Median volume diameters of droplets considered appropriate for various pest control operations by reduced volume application

Median diameter of droplet (µm)	Appropriate use	Remarks
20	Space spraying in enclosed structures	Little, if any, lasting effect
20–60	Contact sprays against flying insects	Low deposition on large obstacles (drift-prone)
60–150	Contact and residual sprays against most pests and diseases	Check volatility; good canopy penetration
150–250	Not normally used in U.L.V. application, but used in L.V. and V.L.V.	Good deposition, but possible loss of cover density at low application rates
250	For larvicide application, bait-spraying, and herbicide application from ground spraying equipment	Good deposition, but with loss of cover density

volume of spray to be applied, not only in run-off spraying but also in reduced volume application.

Among the most important factors that influence the efficacy of reduced volume application is droplet size which determines average cover density. Certain ranges of droplet size have been found most appropriate for particular pest control purposes, as indicated in Table 5 (Anonymous 1980).

2.5.5.2 Uniformity of Coverage

Most application procedures aim at as uniform as possible a deposit over the entire surface treated, i.e. at maximal deposition efficiency. Such efficiency may be defined as that part of the active ingredient emitted which is deposited on the plant surfaces and is retained by them. The efficacy of control derives from only a few percent at best of the total quantity of pesticide distributed.

Pesticide coverage on plant surfaces can be assessed in various ways. One measure is the *area* or the *proportion* of the surfaces covered, namely: (a) whether the pesticide has covered one or both sides of the leaf, and to what extent; and (b) how well the spray has penetrated to the lower layers and the less accessible parts of the plant. Another aspect is the *density* of cover or thickness of deposit, which can sometimes be readily seen, as with wettable powders, but is difficult to quantify unless analytical methods are used. Density of cover relates to the number of particles or drops per unit area on those parts of the plant that have been reached by the pesticide, e.g. 35 drops per cm^{-2}. *Uniformity* of cover is a measure of the densities recorded on various parts of the plant or within the treated swath, while *contact area* means the sum of the area actually covered by all particles or droplets, in relation to the total leaf area.

The result of a spray treatment may therefore be foliage that has its area covered uniformly, but at a high or low density of coverage, which may or may not be adequate to control the target pest. A different type of coverage is achieved by run-off spraying, where the plant surface is to be covered totally, but the volumetric uniformity of the cover is extremely uneven (Fig. 10).

The uniformity of cover is determined both by the characteristics of the treated plant (growth habit, surface properties of leaf and fruit) and the efficiency of the application process. In the treatment of more or less two-dimensional areas, e.g. in placement spraying of soils or very low crops, a fair measure of uniformity can be achieved by the use of multi-nozzled boom sprayers. The degree of variability considered acceptable is a function of the target's size. As measured on targets of about 400 cm^2, fluctuations should not be greater than $\pm$ 20%. However, when cloud spraying is practiced in a tall and dense crop, such as irrigated cotton, this range of $\pm$ 20% is attainable only within the top level of the crop, while the distribution pattern in the whole crop fluctuates much more widely. It is generally accepted that at its middle height such cotton will have no more than 50% of the deposit found on the upper leaves, and the foliage closest to the bottom no more than 25% of that amount. This decrease in droplet density at different crop levels is due to the progressive screening-out of droplets by successive layers of leaves, each layer with its different LAVI val-

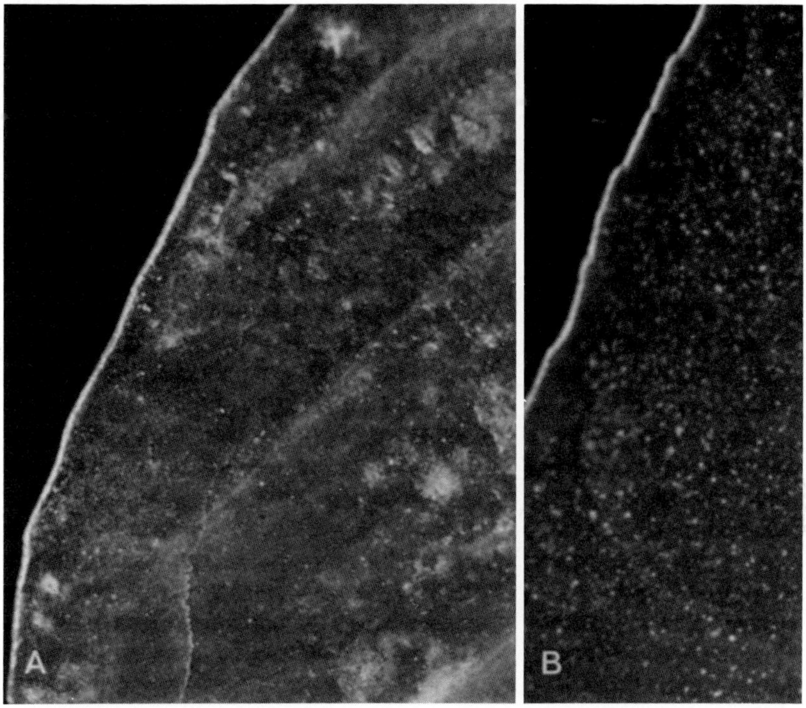

Fig. 10 A, B. Residues obtained by various spraying techniques on citrus leaves. **A** Run-off spraying: uneven distribution of the wettable powder in spots. **B** Reduced volume air-assisted spraying giving a cover density of approx. 250 droplets cm^{-2}

ues. Moreover, since spray drops deposit mainly due to sedimentation or impaction, down-wind leaf surfaces and the under surfaces to leaves are not adequately covered, except under windy or turbulent conditions.

The development of ground sprayers with booms 18 m or more in width, which are often used in cloud spraying of fungicides and insecticides, poses special problems: The booms bounce at their ends and there is considerable lateral whip. These uncontrolled movements, which increase with boom length, cause uneven cover. Special suspension systems to dampen these movements have been developed and consist of pendulum mountings and the use of torsion bars.

The variability of coverage by aircraft application exceeds that obtained in ground application by boom sprayers. Such variability may reach ± 50% (Anonymous 1980), which means that a spray applied at 30 l ha^{-1} will result in crop deposits ranging from 15 to 45 l ha^{-1} at the crop's top level; only 25% of the 15 l may reach the lowest level.

More uniform distribution patterns are most likely to be achieved by air-assisted spraying at reduced volumes, where the air not only carries the droplets to the foliage, but also aids in opening up and agitating dense foliage to facilitate penetration; this also causes coverage of the underside of leaves.

Of all spray techniques, drift spraying has the least uniform distribution pattern. Because of the wide swath, such spraying is often used for the control of flying insects, when it is important to treat large areas rapidly.

As stated above, uniformity of coverage is not always essential for good pest control. But if lack of uniformity is such that deposits in certain parts of the plant are insufficient to kill the pest (e.g. less than 1.5 × LD 99), the result may be the persistence of an endemic population, and subsequent build-up of resistance to those pesticides that have reached the pest in sublethal doses.

As already stated, the mobility of the pest to be controlled, such as a moth or a feeding larva, determines the required cover density (droplets per cm^2). Where protectant fungicides are applied, their area of action exceeds that of the prime deposit for two reasons: (1) as mentioned previously, many fungicides have an area of influence around the actual deposit, and in this area the host tissue is protected; and (2) the protected area increases by the redistribution of chemical deposits due to rain, dew or overhead irrigation, so that the surface of tissue eventually protected may greatly exceed the area of prime contact.

In sum, uniformity of application is not the main criterion by which to judge the efficiency of an application technique. But the more uniform the deposit, the smaller the amount of pesticide that will be wasted, as we choose the dosage which, even at points of least deposit is lethal to the pest.

2.5.6 Dosage and Calibration

Pesticides must be used at their lowest effective doses, because they are expensive, and the higher the dose, the greater the danger to non-target organisms, and the higher the potential residue on food or feed. Dosage rates are stated on manufacturers' labels as percentages, if the material is to be applied by run-off application techniques; they may also be stated in a range of amounts of formulated pesticide per crop area, e.g. $1-2\ kg\ ha^{-1}$, where the lower rate is meant for crops with a low LAI value, and the higher rate for full-grown crops and those with high LAI values. Dosages for reduced volume application are generally stated in such quantitative ranges.

As stated before, the residual deposit from run-off spraying at a given concentration will be almost constant, and application at an excessive rate (overspraying) results in pesticide wastage. In reduced volume application, overspraying has been mentioned to be conducive to phytotoxicity as well as wastage. It should therefore be noted that labels do not always specify the dosage at which materials are to be applied at reduced volume. As a rule of thumb, this dose is usually lower by 30% than that applied by comparable wetting techniques. Thus, if run-off spraying of a potato crop at 0.3% concentration with $1000\ l\ ha^{-1}$ amounts to application of $3\ kg$ pesticide ha^{-1}, an equivalent reduced volume application would use only $2\ kg\ ha^{-1}$; assuming that this application is made at the rate of $100\ l\ ha^{-1}$, a concentration of 2% should then be used.

Application equipment must be calibrated carefully before use. Output rates from dispersal systems, such as nozzles at a given pressure, can be seen from manufacturers' catalogues. The output rate of a spray boom travelling at a given speed can then be calculated.

Preliminary calibration of individual nozzles or spray booms should be carried out while the machine is stationary; output per unit time is thus determined. Final calibration should, however, be made under actual working conditions in the field, with the pesticides to be applied, and with swath width and forward speed as planned.

It may then be found that, although nozzles have been selected for an output of $50\ l\ ha^{-1}$, the actual output in the field comes to $60\ l\ ha^{-1}$. This can be corrected either by altering the tractor speed, but this is not practicable if work is to proceed at the optimal speed feasible for given field conditions; or by decreasing the output rate by changing nozzle tips. The pressure should never be changed greatly (more than 15%), as large changes alter the spray angle and droplet spectrum. This is of particular significance when fan nozzles are used for placement spraying.

Air-assisted spraying, as applied to trees and bushes, requires that, prior to the calibration of liquid output, the optimal forward speed of the machine be determined. This speed is the principal factor in obtaining optimal deposition on distant parts of the foliage.

The sprayer's optimum speed is also determined by considerations of the foliage density, as expressed by the LAVI value of the crop (see Chap. 2.5.1); the greater the density of foliage, the slower the sprayer will have to travel. Then, if radial fan blowers with high velocity airstreams are used when spraying full-grown citrus trees, in which the LAVI of peripheral foliage is high, the speed may have to be limited to $1.25\ km\ h^{-1}$. But for deciduous trees with similar dimensions but more open peripheral foliage, speeds of $4\ km\ h^{-1}$ may be practicable. However, where axial blowers are used on dense foliage crops, such as citrus or grapes, the foliage tends to "close" if the sprayer travels too slowly, and penetration may then be negligible.

The aim in air-assisted orchard spraying is in general to reach target sites in the uppermost part of the tree at an air velocity of not less than $5\ m\ s^{-1}$, as measured about 1 m below the top, in the centre of the tree. This value cannot be calculated and must be measured empirically for each crop. Since air velocity and air volume emitted are constant for a given air sprayer, the air speed at the target can only be adjusted by changing the tractor's forward speed. The sequence of operations is therefore as follows:

1. Measure the air speed at the target site when the sprayer travels at various speeds. This measurement is taken with a ball anemometer, a relatively simple and inexpensive instrument, in the centre of the tree, as described above.
2. The operational tractor speed is that, which has resulted in an air velocity of $5\ m\ s^{-1}$ at the target site.
3. Calculate the rate of liquid output at this speed.
4. When air-assisted run-off sprays (medium volume) are applied, liquid outputs may often be reduced by applying small droplet sprays and by orientating the nozzles so as to permit foliage wetting with minimal liquid output. In

reduced volume spraying, where the deposit cannot easily be evaluated by eye, spray deposit measurement (see Chap. 2.5.5) will aid in the orientation of spray nozzles and thus improve uniformity of coverage in the various parts of the tree.

2.5.7 How to Measure Efficiency of Application

The efficiency of pesticide application may be defined by (a) the quantitative distribution attained, i.e. the amount of pesticide deposited on a given surface area, and (b) by the qualitative distribution, which denotes the type and evenness of deposition. The sampling techniques suitable for this purpose have been discussed in Section 2.4.1.2.

Quantitative distribution is mostly determined by chemical analysis, which requires special facilities and a trained staff to which advisers may not have ready access, and is also time-consuming. Simpler means have therefore been sought to aid advisers and farmers, and the use of tracer materials has become popular, to determine values approximate to those of chemical distribution. The most commonly used tracers are fluorescent materials, which can be recovered from artificial targets, such as slides, or from plant material. However, it must be borne in mind that the tracers do not necessarily behave like agricultural chemicals, and that the efficiency of collection from artificial targets differs from that from leaves or fruits; quantitative values obtained through tracer techniques can therefore only be considered as indicative.

Qualitative distribution relates to the pattern and uniformity of deposition, e.g. the number of droplets or particles per cm^2. This may involve determination of the contact area, which is the total area covered by the deposit on a given site, expressed as percentage of the entire area of that site. Qualitative assessment of distribution is usually visual and therefore inherently subjective and is generally used for the evaluation of reduced volume applications. It has been shown that, with some practice, replicable results can by this technique be obtained with fluorescent materials: droplets as small as 15 μm are "visible" when exposed to ultraviolet or black light for evaluation in situ, and the equipment needed for this purpose is inexpensive (Courshee and Ireson 1961). As many as 1500 samples can be evaluated by one skilled operator in 5–6 h. Another technique consists of the distribution within the crop of water- or oil-sensitive papers, normally applied on each side of the leaf to form a "sandwich" (Anonymous 1981). This makes it possible to distinguish between deposits on the upper and lower leaf surface, but involves a not inconsiderable amount of work.

The techniques briefly referred to above can provide on-the-spot information, or evaluations made within a short space of time, to guide advisers and operators in assessing pesticide distribution. They should be used wherever possible, especially on crops occupying large volumes of space, to indicate to what extent the pesticide has been deposited on critical surfaces.

It must, however, be stressed that distribution constitutes only one of the factors that determine efficacy of pest control by chemicals. Important additional factors, not always sufficiently considered, are the secondary distribution mechanisms of pesticides in the crop, the level of re-distribution of systemic chemicals within the plant, vapour pressure, weather factors affecting the half-life of chemicals, pest behaviour, and the characteristics of the target surface. Since pest behaviour is so obviously of critical importance, bio-assays are a most desirable tool for the evaluation of control efficacy, and facilities for such assays should be fully used wherever possible.

2.5.8 Development of Application Techniques Promises to Raise Efficacy and Minimize Quantity of Pesticides Applied

Modern farming is a production process that has to keep its cost low, yet uses enormous quantities of highly expensive and toxic materials under the eyes of a public acutely aware of the dangers of pesticide pollution. It may safely be stated that few pest control measures, including integrated pest management, have a potential as great as that of advancement in application methods, for reducing the amounts of pesticides the farmer uses. It is therefore regrettable that not enough has been done to improve application methods and techniques in the past.

Pesticide application is an interdisciplinary science, and this fact may to some extent account for its relatively slow development. What has been achieved is largely due to the efforts of interested industries and of public health and vector control services. Interdisciplinary research for application of agricultural pesticides is a fairly new concept. Recent studies of controlled droplet application, the use of air foils, and techniques such as electrostatic spraying, may be a step in the right direction.

There is a vast difference between the theoretically effective dose, as defined in the laboratory, and the field dose recommended to the grower. The latter dose is influenced by numerous considerations including weathering and chemical breakdown (which largely determine the lasting effect) and mechanisms of pest-chemical contact. But the prime factor for field doses being fixed at so high a level is the inadequacy of application techniques and machinery at our disposal for uniformly depositing chemicals on crops with multiplane shapes, multilevel surfaces and highly diverse surface properties. It is widely accepted that with techniques currently used, application efficiency is in the range of 5–6%, but it is often much lower. In an extreme case, that of oil application to citrus to control armoured scales, only 0.5–0.6% of the oil applied is actually retained (Furness et al. 1979).

Successful application is based on knowledge. The best application machinery cannot control a pest unless the proper pesticide is put to timely use. But good application technique can improve pesticide efficacy and greatly reduce application cost, by saving up to 75% in chemicals and minimizing pollution of the environment.

As has repeatedly been stressed in the preceding sections, the processes involved in pesticide application are complex and interacting. Only intensive study will provide the means to keep adviser and farmer informed. Appropriate application techniques are by no means too complicated for reasonably knowledgeable advisers and farmers to comprehend and employ for optimal pest management.

References

Amsden RC (1962) Reducing the evaporation of sprays. Agric Aviat 4:88–93

Anonymous (1980) Guidelines for reduced volume pesticide application. EPPO, Paris

Anonymous (1981) Water-sensitive paper for monitoring the spray distribution. CIBA-GEIGY, Agric Div, Basle, Switzerland

Clifford R, Hart A (1979) Use of the scanning microscope and cathodoluminescence in studying the application of pesticides to plants. Pestic Sci 10:341–357

Courshee RJ, Ireson MJ (1961) Experiments on the subjective assessment of spray deposits. J Agric Eng Res 6:175–182

Davies CN (1966) Deposition from moving aerosols. In: Davies CN (ed) Aerosol science. Academic Press, London New York, pp 393–445

Dimmick RL (1969) Mechanics of aerosols. In: Dimmick RL, Ayers B (eds) An introduction to experimental aerobiology. Wiley-Interscience, New York, pp 3–21

Fraser RP (1958) The fluid kinetics of application of pesticidal chemicals. In: Metcalf RL (ed) Advances in pest control research, vol 2. Wiley-Interscience, New York, pp 1–106

Furness GO, Howard RT, Smith PT (1979) Measurement of petroleum oil deposits on citrus leaves by gas-liquid chromatography. Pestic Sci 10:478–484

Green MB, Hartley GS, West T (1977) Chemicals for crop protection and pest control. Pergamon Press, Oxford New York

Hinds WC (1982) Aerosol technology. Properties, behavior and measurement of airborne particles. Wiley, New York

Johnstone DR (1973) Spreading and retention of agricultural sprays on foliage. In: Wade van Valkenburg (ed) Pesticide formulations. Dekker, New York, pp 344–384

Johnstone DR, Johnstone KA (1977) Aerial spraying of cotton in Swaziland. PANS 23:13–26

Matthews GA (1975) Pesticide application methods. Longman, London New York

May KR, Clifford R (1967) The impaction of aerosol particles on cylinders, spheres, ribbons and discs. Ann Occup Hyg 10:83–95

2.6 The Crop Protection Adviser

2.6.1 The Crop Protection Adviser in Public Employ: Tasks, Limitations, and Contacts

Y. ELKANA, J. PALTI, and R. AUSHER

2.6.1.1 Aims and Tasks of the Adviser

In considering the overall aims of the adviser's work, it must be realized that certain difficulties often arise, both against the background of tasks imposed by his employers and in his relation with growers.

The government's farm policy does not necessarily coincide with the cropping aims of individual farmers. The government may wish to promote certain crops it can market at a profit, or favour growing out-of-season crops for export rather than crops for consumption in the country, although such crops may present particular crop protection problems. If the adviser is seen to push the crops advocated by the government, his advice may lose credibility in the eyes of the farmer.

It follows that the tasks of enforcing the law and of promoting cropping schemes serving government policy should as far as possible be kept separate from advisory work. This is true for all farm advice, but perhaps even more important in crop protection where urgent control operations may fail unless based on factual information only the farmer can furnish.

The crop protection adviser has a wide range of tasks and responsibilities. He should attempt to foresee possible crop afflictions, he has to be able to diagnose their nature as they become apparent, to estimate their future course, and to recommend preventive action or control operations. These recommendations have to be based on damage thresholds, prices of produce, costs of control operations, and on environmental and legal constraints of the proposed treatments. What the adviser requires for this purpose is:

a) Detailed pest and disease control know-how, including methods of diagnosis, prognosis and control.
b) Field experience to assess crop damage apparent and expected; and preferably also.
c) Some economic training to translate expected crop damage and estimated costs of control into a viable appraisal of benefits likely to accrue from the control.

The crop protection adviser thus has to master a considerable range of knowledge and information, which needs constantly to be brought up-to-date. This is one reason why the question arises, whether he should specialize in one of the disciplines of crop protection, i.e. be an extension phytophatologist, entomologist or nematologist dealing with a limited range of parasites and problems in all or most crops in his region, or whether he should preferably aim to be an all-round protection man with adequate knowledge of *all* that could damage a more limited number of crops.

The answer to this question for crop protection advisers employed by public agencies depends on the tasks (additional to farm advice) imposed on the adviser, on his professional proficiency, and on crop and personnel considerations, which will be discussed in this chapter.

An important aspect of advisory work in general, and of crop protection work in particular, is current liaison with legislative, regulatory, research and commercial organizations. Such liaison devolves most commonly on the superior grades of crop protection advisers, as described in the chapters on advisory work in European countries and Israel (Chaps. 3.1 – 3.3).

2.6.1.2 Non-Advisory Tasks Sometimes Imposed on the Crop Protection Adviser

As mentioned above, the urgency often inherent in pest control causes many governments, especially in developing countries, to set up crop protection services before arrangements have been made for general advisory services. Under such conditions, the tasks of the crop protection personnel is apt to include elements of farm instruction as well as policing and supervising implementation of pest control operations. How effective can crop protection advisory work be under such circumstances?

Advice vs. Policing and Policy. More than any other discipline in agriculture, crop protection has numerous legal, regulatory, supervisory and even policing aspects (see Chap. 1.1.2). These may include matters relating to the use of land, water and seeds, the yield of crops (is it free from pests?), the use of pesticides, storage practices, and many other aspects. The primary aim of the agricultural adviser is to help the farmer to increase his income. However important it may be, especially for the longer term, to execute government policy, the farmer's trust can only be acquired if he is convinced that the benefits he can reap from his field are the adviser's chief concern. There are obvious limits to this, and they are most pronounced in the case of crop protection, where one farmer's practices may affect the health of other farmers' crops. But apart from such special cases, it must always be remembered that the basis of all farm advice is accurate information on all that concerns the crop; if the farmer cannot expect such information to be treated confidentially, to be used for advice to his benefit, he may withhold important details and thus impair the effectiveness of advice.

Advice vs. Implementation. The crop protection adviser is likely to be the person most familiar with all practical aspects of pest and disease control in his region.

Authorities therefore often incline to instruct him to put his advice into practice and implement control operations, including:

a) Storage of pesticides and application equipment until required.
b) Allocation and distribution of pesticides and equipment when required.
c) Actual application of the pesticide where required.

There is little to be said against the use of all available personnel where the aim is control of catastrophic infestations, e.g. by locusts. However, where the purpose is routine distribution of pesticides and recurrent control of rodents, birds, bats and other pests to be fought on a regional scale, the crop protection adviser should not be burdened with such tasks. They will take up much of the time he should devote to updating his information and to the field work which is of the essence.

Thus, in the Indian state of Madhya Pradesh, a problem arose when the extension staff were charged with the execution of control operations. When the Training and Visit system of extension was introduced in this state, in 1977, it was decided to include within the extension organization regional crop protection units, consisting of a number of operators with motor sprayers, and a store for pesticides. A conflict of interest then emerged between the "fire-fighting" type of activity, which is of particular importance in emergencies and must then be given every priority, and the on-going regular advisory work of the staff. When this was recognized, the control units were separated from the extension set-up and were transferred to the responsibility of the Agro-Industrial Corporation, which managed the units on a semi-commercial basis. This had benefits for everybody: the farmers got better service, the operators and their equipment functioned in a separate organization coordinated by the district director, and the advisers could do their work in an orderly manner.

2.6.1.3 Allocation of Manpower to Crop Protection Advisory Work

Various aspects of the background required for crop protection advisers, of the tasks that may face them, and of the integration of their work with that of generalist crop advisers have been discussed above. Reviewing these factors, and reiterating some of the considerations put forward in the chapter on the place of crop protection advice in the general scheme of farm advice (Chap. 1.2) we shall attempt to summarize the diverse aspects which have to be borne in mind in allocating manpower to crop protection advisory work, i.e. what determines how broad a range of crops and how many farm units the adviser can be expected to serve satisfactorily.

Table 1 presents an overview of the aspects to be considered for this purpose. The crop protection adviser to which the table refers is not specialized according to disciplines, i.e. he is not an extension phytopathologist, entomologist or nematologist, but is expected to cover all pest and control problems on a more or less limited number of crops. We are quite conscious of the fact that none of the statements made in the table are without exception, and they are presented more to stimulate thought than to lay down rules. The problem of

Table 1. An overview of factors to be considered in the allocation of manpower to crop protection advisory services run by public agencies

Aspects to be considered	Range of crops and problems that can be handled by crop protection advisers
Educational level of farmers	Crop protection advice to farmers at a low level of education and adaptability should preferably be given by generalist crop advisers; where these are not available and the crop protection adviser is called in instead, his specialized knowledge is partly wasted, and this has to be avoided, because the number of such advisers is always limited and their work should be put to better use. One such use is advice to advanced farmers, which often has to include adaptive experimentation and review of up-to-date literature.
Availability of generalist crop advisers	Where generalists are available for cropping advice, they form the chief target group for the crop protection adviser's work; this requires him to maintain close relations with research staff, up-to-date reading, etc. and limits the range of crops with which any one crop protection adviser can deal effectively.
Economic value of the crop and risk of crop loss	Valuable crops need close attention, but are often grown by well educated farmers who can exercise their own judgement. Highly valuable crops with extremely low treatment thresholds are often treated routinely, yet have to be kept under observation to detect occurrence of unusual pests and diseases. The range of high value crops that can be handled by the crop protection adviser is always more restricted than that of crops of lower value.
Crop seasons	Where crop seasons are sharply defined, e.g. by a rainy and a rainless period, crop protection advisers may be able to deal with one set of crops in one season, and with another set in the following season, e.g. with winter cereals first, and then with summer crops (cotton, deciduous fruit). Where crop seasons overlap greatly, the range of crops to be allocated to one adviser will be narrower.
The crop protection adviser's proficiency and capability	This is a decisive factor in the allocation of tasks: experienced, knowledgeable and dynamic advisers can handle a wider range of crops and problems, as well as liaison with regulatory and commercial organizations.
Complexity of control decisions	Complex decisions, especially those concerned with integrated pest control and border-line cases of injury from non-parasitic and parasitic causes, need much consultation and survey. Crop protection advisers who have to make such decisions can only deal with a limited number of crops.
Availability of supporting services (plant clinics, computers, meteorological service, mass media, research)	Where plant clinics are available, the crop protection adviser can cover a wider range of crops and problems. Current supplies of weather data, especially if these can be processed for forecasts, save a great deal of time, which can be devoted to additional crops or farms. Availability of back-up research is essential to enable the adviser to tackle unusual problems with responsibility and confidence. Mass media can provide much support.

Table 1 (continued)

Aspects to be considered	Range of crops and problems that can be handled by crop protection advisers
Size of area, density and accessibility of farms	The smaller the total area to be covered, the denser the farm populations, and the easier access to the farms, the less time and effort is wasted in getting to the farmer; moreover, in dense farm populations there is a better chance to have advice spread among farmers by word of mouth. This is a favourable background to all kinds of advice, including that by crop protection advisers. However, where farm populations are much dispersed and access to farms is more difficult, limitations of time and transport facilities may make it impracticable for more than one adviser to make routine calls, and this should as a rule be the crop generalist.

allocation of manpower is a basic one, and requires constant re-evaluation in the light of changes in the development of farmers, crops and pests, and of the number and proficiency of advisory personnel available.

2.6.1.4 All-Round Crop Protection Advice or Specialization by Disciplines?

We now return to the question, whether the crop protection adviser should be permitted or encouraged to specialize in any one discipline, and become an extension phytopathologist, entomologist, weed control expert or pesticide application specialist. It is perhaps surprising that this problem – one of basic importance in setting up an advisory service in crop protection – has not, as far as we are aware, been discussed more fully in crop protection literature. The decision on this point obviously depends largely on many of the aspects mentioned in Table 1, especially on the target, size of crop loss at risk, and availability of supporting services (see also Chap. 1.2).

2.6.1.5 A Plea for Better Economic Training for Crop Protection Advisers

A considerable amount of space has been devoted in this book to economic considerations relating to the protection of crops, including the assessment of a crop's value, of the damage threshold, and of estimating the amount of preventable damage in order to define rational treatment thresholds (Chap. 2.3). These are complex tasks, and when it comes to crops being threatened by pests, the farmer is often apprehensive and his decision not too rational. As pointed out elsewhere (Chap. 1.4), it is not easy, and sometimes requires some courage, to dissuade a farmer from carrying out a treatment, which, in the adviser's view, is not called for, or could at least be delayed.

These and other decisions relating to the protection of crops require clear economic concepts on the part of the adviser, and this is not always provided in the training of field staff, and perhaps even less in that of laboratory staff. The constantly changing costs of crop protection input (materials, labour, machinery, to mention only direct input) and the changes in the revenue from produce according to market factors, make it absolutely essential that the adviser should have a good grasp of the factors at work, and base his advice on their evaluation. Thus, Thompson and White (1979) have rightly pointed out that more thorough economic training for the adviser (as well as for the grower), is of crucial importance in crop protection, and especially in integrated pest management.

2.6.1.6 Integration of Work of the Crop Protection Adviser and the Research Work

Integration of Advisory and Research Work. Crop protection is a discipline which is complex and ever subject to surprises: new parasites appear and established parasites change the rhythm and severity of their attack, often in response to the introduction of new crops or varieties or to changes in cultural practices, e.g. the introduction of trickle irrigation. Crop protection specialists therefore frequently operate under pressure: crops have to be saved, growers and consumers must be protected from chemical hazards, and surprises must be forestalled as far as possible. What crop protection staff is therefore required to do is to recognize problems in field and plantation without delay, to determine their causes, to search for means of prevention or cure, and to get farmers to apply such means with all reasonable speed. To this end, crop protection advisers need all the help they can get, on the one hand from research, and on the other hand from observant and progressive farmers.

Research personnel can rarely be in day-to-day contact with farmers, and often lack broad practical farming knowhow. Field advisers generally lack laboratory facilities and are not conversant with refined techniques required for more basic investigations, nor do they have time for the latter. In crop protection, therefore, at least as much as in any other sphere of agriculture, progress depends on close cooperation between research and advisory staffs.

Carter (1969) has defined four categories of increasingly complex problems that face the extension worker in developing countries and require increasing degrees of research participation:

a) Causal agent known and control measures possible, e.g. blast (*Pyricularia oryzae*) on rice or Sigatoka disease (*Cercospora musae*) on banana.
b) Causal agent known, but control measures either inadequately known or lacking entirely, e.g. corn borers or virus diseases of rice.
c) Insect pests neither the specific identity nor the control of which are known, and secondary diseases, including certain borers of fruit trees, dieback of mango, and cashew leaf blight (*Phytophthora palmivora).*
d) Diseases of uncertain or unknown etiology, e.g. various diseases of the coconut palm, and cankers of mango.

All plant protection advisers are obviously required to deal with the first category. But handling the remaining categories requires increasingly specialized training and the cooperation of research staff, which in its turn can profit considerably from the adviser's observations in numerous fields or orchards, growing under a variety of conditions. The third and fourth category should essentially be dealt with by research staff.

The extent to which advisers for control of crop pests and diseases need the backing of research staff obviously depends greatly on the level of knowledge and experience of the adviser. In the following we have attempted to generalize and draw up a list of the principal aspects on which advisers need help from research and on which they may be expected to assist research in finding practical solutions to problems on the farm.

Two trends can be recognized: (1) in developed countries, characterized by highly intensive cropping, and complex production and marketing problems, the trend is towards specialization by disciplines and increasing participation of advisers in research, or for "split" extension-research appointments. This trend is evident from the figures presented by Jacobsen (1983) with regard to growth in numbers, from 1953 to 1982, of U.S. extension pathologists involved in research; (2) in developing countries, with lower levels of cropping intensity, simpler production and marketing problems, and little current information from outside agencies (weather, prices), the trend is for all-inclusive crop protection advice, which in some cases even extends to general cropping advice.

The most important argument against specialization of advisers by disciplines is the necessity for integrated pest and disease control, the development of over-all phytosanitary management. The crop protection adviser finds himself increasingly involved with other advisers in devising protection strategies that base on considerations of crop culture, economic thresholds, educational aspects and pesticide application knowhow. He is thus to be seen as a person who is a partner in many teams to which he contributes comprehensive approaches of crop protection. The crop-oriented "integrated" adviser may be expected to furnish broad-based and competent contributions, while advisers specialized by discipline are apt to lose some of the breadth of judgement required for practical advice, even though their knowledge deepens. It must, however, be recognized that the crop-oriented, much more than the discipline-oriented, adviser is in need of the current supply of specific, new information from research specialists.

Another problem, on a more personal level, has been pointed out by Golan (see Chap. 3.3): continuous, all-round crop protection work, mainly on one group of crops in a limited region, tends to become routine work, and does not always afford the regional adviser enough of a challenge and scope for self-expression and professional progress. Since it is so important to attract to this work dynamic, go- ahead staff, this is a problem that deserves serious thought. One way of meeting it is to entrust the regional adviser with some interregional responsibilities for work on a certain crop or pest.

In most countries with advanced agricultural economies, authorities have opted for (or quite often simply inherited from the past) specialization of advisers as extension phytopathologists, entomologists, etc. In the light of recent developments in phytosanitary management, the time may have come for less specialization and more all-round crop protection advice.

What the Adviser Requires from Research

A. Identification

1. Differentiation between non-parasitic disorders and pests or diseases especially where these combine, e.g. in stress situations.
2. Elucidation whether a parasite causes primary or secondary damage.
3. Identification of parasites where this cannot be reliably deducted from the symptoms they cause, e.g. of virus diseases, and of some root troubles.

B. Biology of Parasites

1. Number of parasite generations, their spacing and the length of developmental stages.
2. Mode of survival from season to season.
3. Environmental limitations, especially those due to extreme temperatures and humidities.

C. Control

1. Cultural practices that can take advantage of the pest's or pathogen's weaknesses, especially their environmental limitations.
2. Testing of all aspects of control by pesticides: active material and their formulations, combined use of pesticides, techniques of application, phytotoxic and health hazards.
3. The potential, the means and the modes of application of biological control agents.

What the Research Staff Requires from Advisers

A. Before Research Is Initiated

1. Spotting of afflictions in the field and plantation, with comprehensive and accurate description of the conditions under which it has or has not appeared.
2. Assessment of damage caused, of potential damage, and of worthwhileness of expenditure for control.

B. During and After Research

1. Adaptation of the results of research carried out in one or a few localities to the changing weather, soil and cultural conditions in the fields in other regions.
2. Feed-back on the efficacy of control measures worked out by research, on problems arising in their application, on effects of non-target organisms (e.g. red mite epidemics after insect pest control), on the reaction of crops or cultivars not covered by the research, on the possibilities of combining treatments (e.g. insecticide and herbicide), and on economic results obtained.

Mutual Dependence of Advisory and Research Services. From all that has been said above, it follows that advisory and research services are markedly interdependent. Very real problems arise where few or no advisers are available to serve as the eyes for spotting problems to be solved and provide feed-back when research results are applied in the field. On the other hand, advice cannot be kept up to scratch unless it is constantly nourished by research. Deplorable results have been brought about by too stringent separation between research and advisory work in some developing countries. It is a challenging task for those in charge of the professional aspects of agricultural development to maintain a suitable relation and connection between research and advisory efforts.

References

Carter W (1969) Function and training of the extension worker in crop protection in developing countries. FAO Plant Protect Bull 17:25–27

Jacobsen BJ (1983) Extension plant pathology: challenges and opportunities. Annu Rev Phytopathol 21:137–152

Thompson P, White GB (1979) Linking economic and action thresholds in pest management: implications for advisory services. Staff paper 79-38, Dep Agric Econ, Cornell Univ Agric Exp Stn, 8 pp

2.6.2 Pest and Disease Control Advice Given by Representatives of Pesticide Manufacturers

J. Eshel and J. Palti

2.6.2.1 Background

The vast resources of the pesticide industry enable it to carry out by far the largest part of pesticide research in the world. In addition, pesticide manufacturers have set up advisory and marketing networks to promote the efficient and profitable use of pesticides by growers.

It is difficult to generalize about the advisory work of representatives of these manufacturers, because their staff includes a wide range of personnel: from specialists with excellent academic background and much field experience to field staff trained primarily in the commercial sides of pesticide distribution. But a few general aspects will be mentioned here.

Two principal advantages are enjoyed by advisory organizations working on behalf of pesticide manufacturers, as compared with official advisory agencies:

a) The larger manufacturers can draw on the comprehensive body of information collected while a pesticide is being developed, first in research laboratories and then in world-wide testing. This includes information of the greatest practical importance: the whole range of pests and diseases against which a pesticide acts, and also those against which it fails to act; the range of compatibility of materials, formulated in different ways, with other pesticides or with growth regulators and fertilizers; the potential phytotoxicity of various formulations; and preferred application techniques under specific climatic and soil conditions.

b) Flexibility in the number of field staff and its spheres of activity, a sore point with most more rigidly structured official advisory bodies. If and where conditions of special interest arise, e.g. where an epidemic has to be controlled on a large number of farms, or in an emergency, such as sudden development of pest resistance, manufacturers can shift their advisory work force to meet the new situation effectively. In fact, these advisers are often in a better position than anybody else to spot developing pest or pathogen resistance, and to recommend increased dosages of the pesticide in question, or use of alternative pesticides.

However, from the farmer's point of view, advice from industry representatives sometimes has fairly definite drawbacks:

a) It is not always comprehensive, sometimes far removed from the goal of integrated pest management, and then of doubtful long-range value. But in recent years these representatives have increasingly been involved in the establishment of treatment thresholds, and in thus placing pesticide use on a sound economic basis.

b) Even within the field of pesticide use, the advice of representatives sometimes relates chiefly to those pesticides they wish to market at any given time. Thus, where sale of a herbicide is the order of the day, advice on its use may be highly competent, but questions of disease control on the same farm are left open.

c) The target groups for advice by manufacturers' representatives are obviously those growers who can afford to buy pesticides. This leaves a wide range of growers, especially in developing countries, outside the scope of advice given by these representatives, although these growers may be the ones most in need of advice.

2.6.2.2 Advisory Work in Countries With or Without Operative Pesticide Registration Schemes

In countries with properly functioning pesticide registration or licensing schemes, whether these are obligatory or voluntary (as in Britain), manufacturers have to disclose the principal properties of their pesticides in their label specifications. Once these are approved, they form the framework for pesticide advice by official as well as industry advisers. The labels usually contain a warranty that the pesticide in question is to be used only for the purposes and in the manner specified in the label. This, at least to some extent, absolves manufacturers from liability for damage inadvertently caused by the pesticide.

However, in numerous countries registration schemes either do not exist or fail to be enforced. Here the manufacturer's representative assumes a much heavier formal and informal responsibility. He is often the only, or the most frequently available, source of information, has no official guidance to circumscribe his recommendations, and then has to reconcile his commercial interests with what is best for the farmer. He could also be held legally responsible for any crop or health damage due to pesticides applied according to his recommendation, e.g. where pesticide application has been recommended too close to harvest and the produce carries an unacceptable residue.

It is thus obvious that the advisory functions of the representatives of pesticide manufacturers greatly depend on the formal set-up of crop protection authorities in any given country, and on its actual operation.

2.6.2.3 Technical Information

Growers are supplied by pesticide manufacturers with a wide range of technical information, and this is one of the more important components of the industry's advisory work. Manufacturers have compiled comprehensive, lavishly illustrated handbooks, such as Bayer AG's "Compendium of Plant Protection" or the Technical Monographs published by Ciba-Geigy; these are most helpful to growers and advisers, whether official or otherwise. Recently, visually excellent pamphlets on methods of spraying and maintenance of equipment have also been produced by manufacturers of pesticides and application machinery. One advantage of this high class publication effort is that the information is made available to growers free of charge, and that their illustrations are most useful for instruction of semi-literate or illiterate farmers.

Many manufacturers have published pest, disease and weed identification manuals for specific crops, with illustrations of a standard which official bodies find difficult to match; a good example is the cotton pest identification manual issued by Imperial Chemical Industries. Instructive film and video material on pests and their control has been produced by several manufacturers in recent years.

The major part of all information available on toxicity of pesticides to man, livestock and wildlife is the fruit of research carried out by pesticide manufacturers. Official instructions to growers regarding health hazards involved in the use of pesticides admitted for registration are largely based on this information.

Most large manufacturers have concentrated all the technical information on their pesticides in computerized data banks, which cover a very wide spectrum: pesticidal and phytotoxic properties, compatibility and effects of soil conditions, health hazards and antidotes etc. This information can sometimes solve, or at least clarify, problems that arise locally in the use of pesticides, and this is of paramount importance in countries that have no crop protection research of their own.

2.6.2.4 Field Tests and Demonstrations

Since the performance of every pesticide depends greatly on actual field conditions, including soil type, irrigation and other practices, great importance attaches to local tests. These are widely carried out by manufacturers' representatives, with or without the cooperation of crop advisers or crop protection officials. Such tests may serve to furnish evidence required for pesticide registration. But where such registration is not demanded, they may be the best source of detailed definition of the optimal conditions for a pesticide's use.

One of the most convincing ways of showing the benefits of pesticides and of their modes of application are field demonstrations. These are most useful for showing the behaviour of pesticides under actual field conditions. Manufacturers' representatives effectively employ this technique to familiarize growers with recent developments in pest control by chemicals. The spectacular differences between treated and untreated plots which often obtain in such demonstrations are equally impressive to progressive and illiterate farmers.

2.6.2.5 Current Advice and Timing of Pesticide Use

As pointed out above, one of the advantages of advisory services run by manufacturers is its flexibility. This may find expression, where deemed necessary, in frequent visits to a given farm, to determine when pesticides should be applied with optimal effect or to check on the level of their efficacy where it is suspected that pests or pathogens may have become resistant. This, in fact, is monitoring, something the official crop protection adviser rarely can afford the time to do. Insect trapping and other means of verifying pest activity are also used by the representatives. Where data banks for pest management exist, the staff of manufacturers can contribute data and widen the circle of growers who profit from such schemes.

Among the recent developments for optimal timing of pesticide use are the pocket computer programmes, such as those introduced in Britain by Ciba-Geigy Ltd. for the improvement of cereal disease control. One of these programmes, called Spraytel, works as follows: by pressing appropriate keys on the computer, the grower indicates whether the crop in question is wheat or barley, and what growth stage (on a decimal scale provided) the crop has reached; he then answers a series of questions on diseases present or likely to appear, and on weather conditions, soil type etc. When this has been done, the computer will tell the farmer whether or not the information supplied justifies application of a fungicide, and at what dosage.

2.6.2.6 Commercial Advice

Advice given by official advisers sometimes is lacking in practical value, as it may fail to give up-to-date information what pesticides are actually available in a given locality, and at what price. In this respect representatives of pesticide manufacturers are in a much better position to advise farmers which pesticides can be delivered promptly and which may be expected to run short, how various materials and formulations compare in price, and also how best to store pesticides left over from one season to the next. This is practical advice of the kind the farmers value highly.

In fact, the representatives here fulfil a logistic as well as an advisory function. It is their business to see to it that the proper pesticides are available when needed, and that the quantities available meet the requirements to be expected according to crop acreage, dynamics of pest development, and profitability of the crop, and also that the price at which they offer a pesticide is competitive. How important this is to the farmer will be appreciated by any adviser, official or not, who has seen his good advice frustrated by the lack of stores at the critical moment.

A service supplied by some manufacturers is to test, free of charge, the fitness for use of pesticides which have been in storage for extended periods.

2.6.2.7 Outlook

The relationships between official crop protection advice, with its built-in, chronic lack of resources, and industry-given advice, with its obvious limitations, are discussed in Chapter 1.2.7. Coordination of the advice given from these two sources is of extreme importance.

In Chapter 1.4, which deals with the farmers' perceptions of pest control problems and operations, it was concluded that for the foreseeable future the role of pesticides in pest and disease control will gain in importance and will become increasingly prominent, not only in developing countries where pesticides often provide easy solutions (if the farmer has money to buy them), but also in advanced farming, where exclusion of risk is a dominant consideration. The relative importance of industry-based advice may thus be expected to increase, and the introduction of devices, such as Spraytel computers, will help them to do so.

This development should not be in conflict with the goals of integrated pest management. But the manufacturer's representative may generally tend to advise pesticide use even where the economic justification may be in some doubt. He is bound to do so, because he cannot risk crop failure, which would cost him his client's trust. And this trust, perhaps even more than in the case of official advisers, is his principal asset and must be safeguarded if his work is to be effective.

In recent years, large manufacturers of pesticides have begun to contribute studies and advice to two major aspects of crop protection: application techniques and avoidance of environmental hazards, especially air and soil pollution by pesticides.

It is being realized, somewhat belatedly, that evolution and extension work in the economic and effective application of pesticides is the prerequisite to the practical use of these materials where their use has so far been minimal, and handsome leaflets have been issued by several manufacturers with graphic illustrations intelligible to all.

The problems of environmental hazards have become so important to the public image of pesticide manufacturers, that they are most likely to increase their activity in this sphere, and public bodies as well as farmers may hope for relevant information how to reduce such hazards.

Part 3 How Crop Protection Advice Is Given in Various Countries: a Diversity of Objectives and Approaches

In the following seven chapters a description is given of the way advisory work in crop protection proceeds in various countries. This is far from being a systematic review of such work, and includes only a number of examples drawn from countries with farm economies developed in different directions and to various levels, with diverse forms of organization.

In conclusion (Chap. 3.8), some comparisons are made between various types of organizational set-up, and some of the aspects are discussed that require special attention to further crop protection advice in advanced and in developing farm economies with stress laid on the benefits of integrated pest management.

3.1 Crop Protection: The Role of the Agricultural Development and Advisory Service in England and Wales[1]

J. T. FLETCHER

The management of pests, pathogens and weeds is one of the most important aspects of crop production. With continuously increasing costs, it is important to determine the optimum input into crop protection to achieve the most cost-effective use of resources. Farmers in England and Wales require an economic return from their land in order to make a profit. This does not necessarily mean achieving the maximum yield per hectare, but that which gives maximum profitability. At present, arable farming in England and Wales is geared to high output per hectare and this is closely linked with profitability. Changes in the economic pattern may result in farmers striving to reduce costs whilst retaining good yields of high quality produce. In such situations less costly approaches to crop protection are required. Each situation and subsequent change provides new challenges for the farmers, who require the answers to new problems as they occur. These answers are not always available. When they are, the advisory services have a role to play in keeping farmers informed of the latest developments. When they are not, investigations must be initiated to solve outstanding problems. These roles in England and Wales are those of the Agricultural Development and Advisory Service (ADAS) which is a part of the Ministry of Agriculture, Fisheries and Food (MAFF), but also of various other advisory organizations run by chemical companies, agricultural merchants, or by consultants. Each has a vital part to perform if the industry is to remain efficient and capable of responding to rapid change. In this review of crop protection in ADAS the emphasis is placed on the control of pests, diseases and weeds and of the advisory approach taken largely regionally to counteract problems that arise.

3.1.1 The Agricultural Industry in England and Wales

Most farms in England and Wales are highly mechanized. The average farm size is about 60 ha, but most production is from the larger farms. In many parts of England there are farms larger than 400 ha. The farming of crops and animals is distributed over the whole of the two countries, but there are some concentrations of crops in certain areas (Table 1). For instance, the Vale of

1 The views expressed in this article are those of the author and do not represent MAFF policy

Table 1. Regional and national laboratories in ADAS and the main crops in each area

Location	Area of work
Northern Region	
– Leeds	Cereals, potatoes, cucumbers, lettuce, brassicas
= Newcastle	Cereals, potatoes, swedes
Midlands and Western Region	
– Wolverhampton	Cereals, fruit, hops, vegetables
= Shardlow	Cereals, potatoes, vegetables
≡ Kirton	Bulbs, potatoes, vegetables, outdoor flowers
≡ Evesham	Protected vegetables, field vegetables, apples, soft fruit
Eastern Region	
– Cambridge	Cereals, oilseed rape, potatoes, vegetables, fruit, flowers, protected crops, mushrooms
≡ Cheshunt	Protected vegetables and flowers, mushrooms
South East Region	
– Reading	Cereals, oilseed rape, vegetables, trees and shrubs, soft fruit
= Wye	Apples and pears, strawberries, hops, protected vegetables and flowers, mushrooms, field vegetables, potatoes, oilseed rape and cereals
South West Region	
– Bristol	Cereals, vegetables, protected vegetables
= Starcross	Vegetables, outdoor flowers, bulbs
Wales	
– Aberystwyth	Cereals, early potatoes
= Bangor	Miscellaneous horticultural crops
= Cardiff	Protected vegetables, field vegetables, potatoes, cereals
+ Harpenden Laboratory – Harpenden, Herts	All matters concerned with legislation in connection with pests and diseases, import and export, certification schemes. Pesticide residues and pesticide usage surveys. Crop loss assessment; pesticide registration.
+ Slough Laboratory – Slough, Berkshire	Control of insects, mites and fungi associated with stored products
+ Tolworth Laboratory – Tolworth, Middlesex	Control of rodents, wildlife conservation in relation to pesticide residues
+ Worplesdon Laboratory – Worplesdon, Surrey	Biology and control of mammals other than rodents

Key: – Regional centre
= Regional subcentre
≡ Outstationed laboratory
+ Control laboratories
Plant Clinics are situated at all regional centres, subcentres and outstations

Evesham in the Midlands is a centre for field-grown vegetables, protected vegetables and fruit. East Anglia is the major area for the extensive production of cereals, whereas the south-east is the main fruit-growing area. In each area a specialized knowledge of crop protection is required to tackle the problems associated with the crops that are grown.

The educational level of farmers, like that of any other section of the community, is variable. Many farmers learn their trade by working on farms (often belonging to their parents). Most will have received a basic level of education up to the age of 16 years. Some farmers follow courses of further education, and diploma and certificate courses at local agricultural colleges meet many of the needs of the farming community. A minority of farmers have obtained degrees at one or other of the agricultural universities.

3.1.2 The Agricultural Advisory Service

The National Advisory Service. An Act of Parliament in 1890 provided for technical instruction in agriculture and this Act was the beginning of technical advice to farmers. The 1944 Miscellaneous Provisions Act gave powers for the setting up of an advisory service in England and Wales (Scotland and Northern Ireland were provided for separately) and in 1946 the National Agricultural Advisory Service (NAAS) was formed. It aimed to cover all aspects of agriculture including crop protection. England was divided into seven provinces with Wales as an eighth. Later these provinces were renamed Regions. NAAS operated for 25 years and in 1971 underwent a major reorganisation and was combined with other services within a new organization, the Agricultural Development and Advisory Service (ADAS). This reorganization also heralded a change of emphasis of work towards the development of ideas and new techniques and away from the individual contact of advisers with individual farmers and growers. Recently there has been a redefinition of the Regions and there are now six, with six Sub Centres.

The Present Structure of the Agricultural Development and Advisory Service. There are approximately 5000 members of ADAS divided in four major services – Veterinary, Agriculture, Agricultural Science and Land and Water Service (Fig. 1). Crop protection work is done by two services, Agricultural Science for all aspects of pests and diseases, and Agriculture for weeds and the interaction of pest and disease control and other agronomic practises. Crop protection is not therefore dealt with by one unit but by an amalgam of separate disciplines. In addition, there is a separate Plant Health and Seeds Inspectorate, which is concerned, among other things, with the import and export of plant material, the control of pests and diseases by legislative means and the operation of certification schemes for planting material (see crop protection by legislation). Other disciplines with some involvement in crop protection within the Agricultural Science Service are Microbiology, Analytical Chemistry and Soil Science, and within the Agricultural Service, Mechanization, which is partially concerned with the efficient application of pesticides.

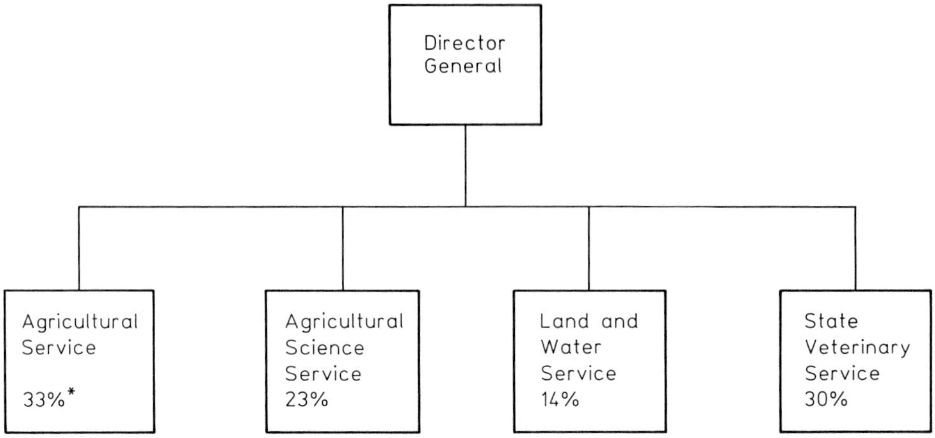

Fig. 1. The structure of ADAS. * Percentage of the total staff number of ADAS

The crop protectionists are situated at either National or Regional laboratories (Table 1). Agronomists in the Agricultural Service have no laboratory facilities, but within the Agriculture Service there are 13 Experimental Husbandry Farms (EHF) and 7 Experimental Horticulture Stations (EHS). Collaborative work is done by members of both the Services at these centres.

There are a total of approximately 140 plant pathologists and entomologists in the Science Service (approx. half of each, of whom some 50 are in the Harpenden Laboratory and 90 in the Regions). There are two national specialist agronomists for advice on herbicides, one for agricultural and the other for horticultural crops, respectively. In addition there are 35 specialist agronomists at Regional Centres and Sub-Centres dealing with all aspects of the agronomy of agricultural crops including herbicide usage. Advice on crop protection is also given by the agricultural and horticultural advisers in the various divisions and districts throughout England and Wales.

The integration of crop protection advice is done partly by the district adviser and partly by consultation between the various specialists. The district adviser is able to gather information from the various specialists and put this together into a plan of action for the farmer. The overall planning of crop protection strategies including the integrated approach to pest, disease and weed control is done at both National and Regional levels through meetings of specialists and by personal contact between the different specialists working with the same crops. Advice on the integrated approach plays a major part in the strategy used for the control of pests and diseases of protected crops, fruit and cereals.

The Wildlife and Storage Biology Discipline within the Agricultural Science Service plays an important role within the general context of crop protection. Advisers in this discipline work at the central laboratories (see Table 1) and within the Regions. Their role is to give advice and do experimental work on rodents and larger animals, e.g. rabbits, birds and pigeons, as causes of crop loss. The discipline is also responsible for the loss of products in store resulting from biodegradation. An important aspect of the work of this discipline is conservation, ensuring farmers take full account of the increased protection given to wild animals and birds by the Wildlife and Countryside Act 1981.

3.1.3 The Communication and Dissemination Pathways

From Farmer to Adviser – The Communication Pathway. The first point of contact for all farmers is the district adviser who is a generalist but tends to specialize in arable or horticultural crops or in livestock husbandry. In most districts of the country there is one agricultural and one horticultural district adviser and sometimes more than one of each in an office covering several districts. Districts are grouped into Divisions and Divisions into Regions. There are 24 Divisions in England and Wales. The district adviser, generally a graduate in his subject, is able to deal with most of the everyday enquiries concerned with crop protection, but when he needs specialist advice he consults his Regional team or sends material to the Plant Clinic. The Regional specialist may visit the farmer together with the district adviser or the district adviser may convey the Clinic diagnosis to the farmer together with the appropriate control measures which, when necessary, he has discussed with the Regional Specialist. The farmer may also consult directly various other government sources of information when making a decision, including the Regional specialists and research workers in the Agricultural Food and Research Service (Fig. 2). This form of contact is considered to be undesirable because it does not allow the most effective use of staff resources. There are, however, circumstances where, by arrangement, it can be very useful.

In addition, there are commercial firms and private consultants who supply valuable information to the farmer on all aspects of crop protection. These advisers are more numerous than ADAS advisers, and provide a great deal of the on-farm advice given to the industry.

Crop protection advice within ADAS operates at two levels, firstly through the generalist in the district, where it is one of his many subject areas on which he advises, and secondly through the Regional specialists for plant pathology, entomology and agronomy (herbicide science). There are no general or specialist crop protection advisers whose sole job it is to advise exclusively on all aspects of crop protection.

From Adviser to Farmer – The Dissemination Pathway. Crop protection advisers contact farmers in various ways, using all the channels of communication available to all advisers, including the spoken word – in direct contact with farmers – and more rarely by mass media. Moreover, within ADAS the Plant Clinics (Table 1) are one of the main contact points.

Communication – The Spoken Word. Crop protection advice is communicated directly to farmers on an individual basis by direct face-to-face contact, by telephone and to large and small groups. Visits to farmers by crop protection

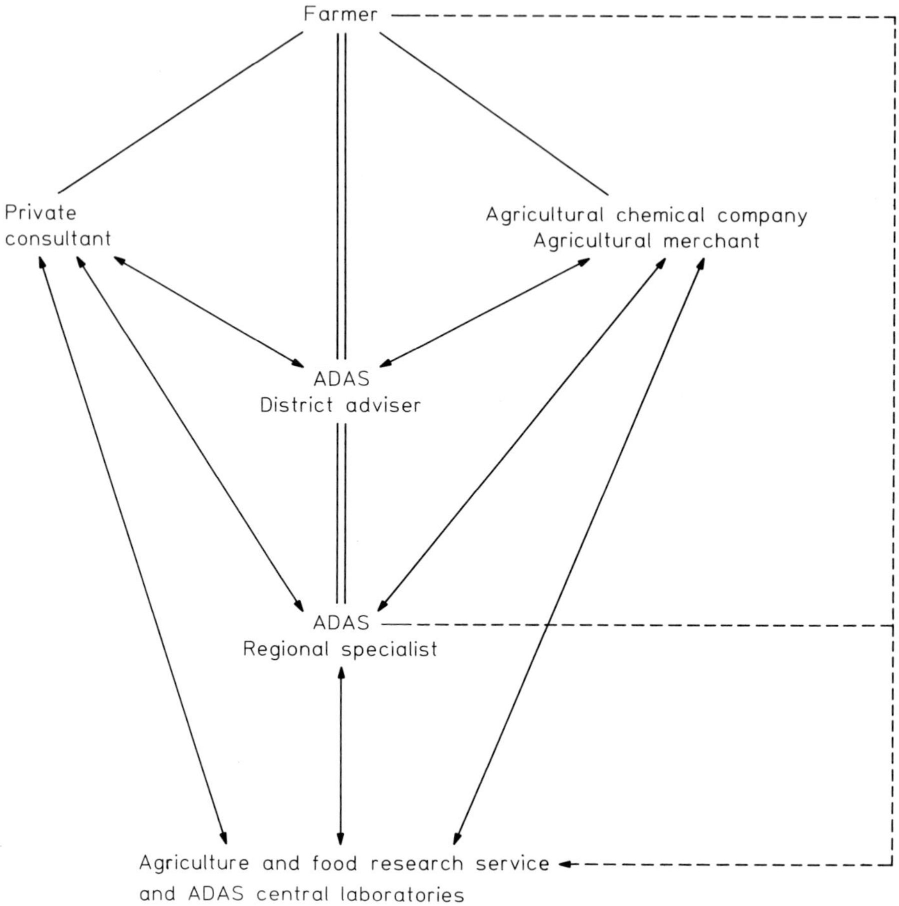

Fig. 2. Communication pathways for crop protection information to and from the farmer. ═ Preferred communication pathway ⟷ Information exchange in both directions. - - - Least preferred pathway

specialists are made almost always with district advisers and occasionally, in addition, with colleagues from other disciplines such as Soil Science. Visits by specialists in crop protection subjects are never made as a routine but only in response to a request to investigate a particular problem. Similar advice is given directly over the telephone, most frequently to district advisers.

The district adviser is also the key organizer of meetings, whether small discussion groups or day symposia. Often small groups of farmers specializing in particular crops meet in turn on each others farms to look at techniques or specific problems of the farm. Crop protection subjects fit well into such discussion sessions, especially for such crops as wheat and barley during the times of the year critical for the applications of pesticides. Occasionally, plant pathologists and entomologists organize information sessions for farmers and growers, where individual disease and pest problems are considered in depth and the latest research results are made known to the farmers.

The least common use of the spoken word is through the mass media or radio and TV. Occasionally broadcasts are made, particularly if there is a development of Regional or National significance.

Communication – The Written Word[1]. Decreasing resources and increasing commitments have resulted in the written word becoming perhaps the main means of communication. This written information is distributed in various forms.

a) Leaflets – these deal with individual pests, diseases or weeds separately; they are usually four or five pages long with one to three coloured pictures. They describe the biology of the organism, as well as the diagnostic features and means of control. Control sections deal with principles as well as specific items. A complete list of registered pesticides is not usually included, because such leaflets are revised at infrequent and irregular intervals and not generally more than once in 3 years. These leaflets are freely available to anyone who asks.
b) Awareness leaflets – introduced in the past 5 years these deal with new or unusual problems, e.g. *Calyptella* root rot of tomatoes, barley yellow mosaic virus. They usually fold out to make one sheet with coloured pictures of symptoms with some basic notes on identification and how to deal with the problem if it is found. These leaflets are usually not revised, as they serve an awareness purpose only.
c) Booklets – cover specific subjects such as pests and diseases of tomatoes or weeds in cereals. They contain a small amount of information on the biology of the organisms, but a detailed account of control measures covering such aspects for pests and diseases as cultivar resistance, cultural control and chemical control measures. They have tables of all the Approved Products for any crop – and also all firms label recommendations for non-Approved Products. There are no illustrations in these booklets. The booklets are revised regularly usually every 1 to 3 years.
d) Reference books – there are a few detailed books covering some crop protection subjects, e.g. diseases of bulbs, diseases of potatoes, which contain pictures of symptoms, details of the biology of the diseases, an outline of the established control measures with references. They deal in greater depth with the subject than other publications and are revised very infrequently. These publications are sold through the Stationary Office bookshops or from the MAFF publications department.
e) News sheets – most disciplines distribute free of charge regular news sheets on the current situation. For instance, Pest and Disease Intelligence is sent weekly to anyone requiring it, as well as to the local papers, radio and TV. It also forms the basis of telephone information service messages which are used in many Divisional offices by district advisers.

1 All ADAS publications including a list of the publications available are obtainable from Ministry of Agriculture, Fisheries and Food (Publications), Lion House, Willowburn Trading Estate, Alnwick, Northumberland, NE66 2PF

A national news sheet of pest and disease intelligence is distributed weekly throughout the summer and at less regular intervals at other times of the year. Reports from all Regions are co-ordinated at ADAS, Bristol, and this information is distributed back to the Regions, but most importantly to the national press and to radio and television, where it gets regular coverage.

f) Articles – are written for many farming journals as well as for ADAS divisional publications. Most divisions distribute a booklet of topical news and articles at monthly intervals and crop protection subjects are regular features in these publications.

g) Scientific Papers – when results of research work have been completed they are reported in the scientific press. Much of ADAS crop protection work is not suitable for such journals, but there are between 10–20 papers a year published from the entomology and plant pathology disciplines.

3.1.4 Plant Clinics

ADAS Plant Clinics are situated at 15 laboratories (Table 1). A Plant Clinic is an essential reference point for the industry and is used by ADAS advisers, advisers from private and chemical companies and by consultants. Its main function is to provide an accurate and fast diagnostic service and to combine with this recommendations which the farmer can use to prevent crop loss. Within ADAS about 25,000 samples are examined annually within the Plant Clinics. The service is free and is available to all commercial farmers and growers. The complexity of the Clinic and its size varies with the Region. At its most developed, the Clinic is staffed daily by a representative from the crop protection disciplines, supported by additional expertise from other disciplines such as Soil Science, Microbiology (for the identification of bacterial pathogens) and Analytical Chemistry. All problems of crop growth are dealt with, and the appropriate department makes any detailed studies that are necessary following the preliminary diagnosis. At the end of the examination the results of all investigations are synthesized into the final report which is sent via the district adviser to the farmer or directly to the farmer, depending upon who made the initial enquiry. Speed and accuracy are the essential ingredients of meeting the farmers requirement. It is no good having a very precise diagnosis of a problem 2 weeks after the need to spray to save a crop. The speed of diagnosis of virus diseases has been overcome as far as is possible by the development of a Virology Branch with small units at Cambridge and Harpenden. Electron microscopy and serology at both centres provide a rapid diagnosis of those viruses that can be identified in this way.

Smaller Clinics are often the responsibility of one department only, usually Plant Pathology, who serve as the centre for enquiries consulting the specialists of the other departments whenever necessary.

The Use of Plant Clinic Data. The Clinics do not receive a random selection of samples from all of the crops of the Region but a selected sample from those crops with problems. They cannot therefore be reliably used for disease intelligence purposes because of the bias of the samples. What is more, the most common pest and disease problems are diagnosed on site by the district advisers and therefore never reach the Clinic. Plant Clinic records will be weak on such diseases as potato blight, barley mildew, or apple scab, although these may be the commonest diseases of the Region. Disease distribution maps cannot therefore be produced from Plant Clinic data for this very reason. However, the new or unusual inevitably reaches the Clinic, which is a good reference point for new pests and pathogens, new symptoms, the breakdown of resistant cultivars and pesticide resistant forms.

Computerization of Plant Clinic Data. Preliminary work on the computerization of Plant Clinic data within ADAS has only recently begun with the intention of replacing much of the paper work involved and providing a more effective transmission of information to the field advisers. The system so far evolved provides little more than an alternative to a day book. It has the advantage that analytical data is easy to obtain, but the limitations of this sort of data have already been pointed out. There is the potential to combine diagnosis with stored data on control, making replies quicker and more effective, and also for the exchange of data between clinics through a network of linked computers.

3.1.5 Solving Problems

A very important role in crop protection for ADAS is the solution of problems where answers do not exist. Such research and development (R & D) is a large component of the work of crop protectionists in ADAS. On average, half of the total time is spent doing R & D. Advisers become aware of problems in various ways, e.g. through the Clinic, on advisory visits, or during informal conversations with farmers. The resources of the disciplines involved in crop protection are managed nationally for R & D, and decisions on work priorities are taken by the head of each discipline within guidelines laid down by an Agricultural Science Service headquarters R & D committee; this is answerable to an ADAS central R & D committee and ultimately to the Chief Scientist (MAFF). For instance, within the plant pathology discipline there are commodity groups for cereals, arable crops, vegetables, fruit, hardy ornamentals, protected crops and grass. Each group consists of about six advisory plant pathologists and the chairman of each group is a Regional Pathologist. Having decided priorities within the group, the head of discipline in conference with the Regional Pathologists decide on the national priorities for R & D within the Regions for all the commodities. Also within each Region the Regional Agricultural Scientist has overall responsibility for the execution of the R & D programme. The Regional Pathologist may add some specifically regional problems to the list of R & D priorities so that each regional programme is a combination of national and regional work. A similar structure exists for the entomology and agronomy disciplines.

Most of the regional R & D work is directly linked to control in one form or another. Commonly pesticides are compared for their efficiency against pests and pathogens, optimum spray timings established and spray equipment compared. Cultivar resistance may be assessed in the field especially if there are indications of the origin of new pathotypes. Work is done on pesticide resistance, on the biology and epidemiology of the organisms causing problems, and on epidemic development, particularly in relation to the environment. Attempts are made to devise forecasting schemes to optimize the use of pesticides. By having a national organization, it is possible to put large resources into specific problems at short notice and to do extensive countrywide work in order to obtain a lot of information in a short time. A recent good example of such work was the survey done in 1983 of resistance of the eyespot fungus (*Pseudocercosporella herpotrichoides*) to the carbendazim fungicides. Over 3300 isolates of this pathogen from 473 different crops were examined for fungicide resistance. Other survey work on foliar diseases of cereals has been done for many years and has provided invaluable information on disease incidence during a period when fungicide usage has risen dramatically. Now most cereal crops get at least one fungicidal application and more often two or more.

Crop protection has featured in some Developmental Farm Projects. Developmental farms are commercial farms where ADAS applies modern techniques and has a side-by-side comparison with the existing standard commercial techniques. In this way farmers can be shown how improvements can be made or different and sometimes less costly approaches taken without loss of efficiency. One such project involved the integrated control of pests and diseases of apples.

The majority of the regional R & D programme in crop protection is done either at a laboratory centre or on farms in the locality. ADAS experimental farms or horticultural stations are also used as sites for regional and national experiments. Most regional centres have glasshouse and plot facilities, but rely very heavily on the goodwill of the farming community for field experiments particularly with crops where large plots are required in order to get reproducible results.

3.1.6 Specialized Branches and Services

There are a number of specialised branches and services within ADAS dealing directly or indirectly with various aspects of crop protection. Their services are available for crop protection advisers within ADAS and also in some cases these units deal directly with the industry.

Pesticide Registration and Approval. Two schemes are operated at the Harpenden Laboratory. The Pesticides Safety Precautions Scheme is concerned with the safe use of pesticides, taking into account the possible effects on human beings, livestock, domestic animals, beneficial insects, wildlife and the environment in general. The scheme has resulted from negotiations with various commercial bodies, including the British Agrichemicals Association, with various government departments, including the Ministry of Agriculture, Fisheries and Food. This scheme applies not only to England and Wales but also to Scotland, Northern Ireland and the Isle of Man.

Under the scheme, clearance is given for the use of pesticides when sufficient information has been presented to satisfy the requirements of the scheme. Initially clearance is given for small scale trials and eventually, by a staged system, full-scale clearance can result. Although this scheme is not backed by legislation at present there are a number of statutes which directly impinge upon it, e.g. Health and Safety at Work Act 1974, Health and Safety (Agriculture Poisonous Substances) Regulations 1975 and the Poisons Act 1972.

The second scheme concerned with registration is the Agricultural Chemical Approval Scheme, which is entirely voluntary. This covers herbicides, fungicides and insecticides and operates in England and Wales, Scotland, Northern Ireland, the Isle of Man and the Channel Islands. It is concerned with the biological efficiency of pesticides to enable users to select the most appropriate crop protection chemicals for particular situations. Products can be "Approved" only after they have been cleared under the Pesticides Safety Precaution Scheme. Approval is granted for specific uses and any change in formulation or method of application even of the same product requires clearance and fresh evidence for approval.

The National Pesticides Residues Unit at the Cambridge Regional Office with a subcentre at Wye is largely concerned with the effects of pesticide residues on crop growth. Most of the work is concerned with the investigation of problems that have arisen largely with herbicide use, although some work is done on insecticides and fungicides. This is in addition to much work done on pesticide residues and pesticide chemistry at the central laboratories (see Table 1).

The Aerial Photography Unit at Cambridge developed from the Plant Pathology Department where it was set up to track the initial foci of potato blight in the Fenlands. Since then it has become a separate unit and has widened its scope to study a variety of problems where an aerial view can give information not obtainable on the ground. The work done on barley yellow dwarf virus, halo blight of beans (*Pseudomonas phaseolicola*) and various soil problems has highlighted the value of this unit to ADAS. Its services are available to all members of ADAS.

The Disease Assessment Branch and the Pest Control and Pest Intelligence Branch are at the Harpenden Laboratory. Here attempts are made to quantify the effects of pests and pathogens on crops and to predict crop loss from damage levels. It follows that such work is closely linked with the biology of the causal organisms and much work is done on the influence of the environment on the development of damaging levels of pests and diseases and on forecasting pest and disease incidence. These branches co-operate with Regional specialists but generally work on their own programmes. A current example of co-operation between the Disease Assessment Branch and the Regional Pathologists is the development of a crop disease environmental monitor (CDEM) which, it is hoped, will enable more reliable disease forecasting procedures to be developed.

Forecasting of pests and diseases is done within the Regions and together with the Harpenden Laboratory new methods are developed. Within the Regions, warnings of outbreaks of various pests are given, including pea moth, cabbage root fly and various fruit moths and aphids (see also Chap. 2.4.3.2 on the trapping of aphids). Light, pheromone, sticky, and suction traps are used to enable this work to be done. Disease forecasts are issued for potato blight (*Phytophthora infestans*), apple scab (*Venturia inaequalis*), fireblight (*Erwinia amylovora*) and hop downy mildew (*Pseudoperonospora humuli*). These are dependent upon the recording of meteorological data from a national network of weather stations.

The Virology Branch has units at the Harpenden Laboratory and Cambridge. Its work in virus disease diagnosis has already been mentioned under Plant Clinics. The work done by this branch is largely concerned with the identification of viruses and mycoplasms. New techniques have been developed for specific purposes. Some epidemiological work is also done.

A small *Mushroom Pest and Disease Unit* is situated at the Lee Valley EHS. Its function is to investigate pest and disease problems of the mushroom crop and to provide information which will allow more effective advice to be given.

Tests of Soils and Propagating Material. Although the majority of services offered by ADAS are free there are some where a charge is made. In crop protection these include soil tests for nematodes, soil tests for various pests, soil tests for apple replant disease, potato tuber tests for virus status and mushroom sporophore examination by electron microscopy for virus content. Soil analysis in conjunction with problem investigation is free but the routine analysis of soil for nutritional status is not.

3.1.7 Crop Protection by Legislation

The Ministry of Agriculture, Fisheries and Food is responsible for legislative crop protection through the Plant Health Administration Unit which decides matters of policy. The Plant Health and Seed Inspectorate who are part of the Agricultural Science Service, carry out the policies. There are 86 Plant Health and Seeds Inspectors (PHSI) situated in various Divisional offices, at Sub Centres and at Regional Centres. The headquarters laboratory giving scientific support to the PHSI is the Harpenden Laboratory where facilities exist for work on nonindigenous pests and diseases. The Regional laboratories are sometimes also used to help with diagnostic work.

The main work of the PHSI is concerned with the prevention of the introduction of non-indigenous pests and diseases, the inspection of plant material for export to enable international plant health regulations to be maintained, to organise eradication of pests and diseases that are accidentally introduced, to maintain the health standards of certified stocks of some crops, e.g. potatoes, and to execute legislation designed to contain particular problems on hop wilt and fireblight.

3.1.8 Management of Crop Protection in ADAS

The headquarters management team in London, headed by the Director General who is assisted by the Heads of Service, determines the overall strategy for all aspects of the work and management of ADAS. The strategy takes into account government policy as well as the views of the industry and of ADAS. Priority areas are determined and ADAS works within these. The farmer's view is obtained through Regional committees known as the Regional Panels where 12–14 farmers representing the local industry debate with ADAS the needs of the industry and the role of ADAS in meeting those needs. Farmers are also represented on some R & D committees. The ultimate responsibility for all the activities within ADAS rests with the Director General.

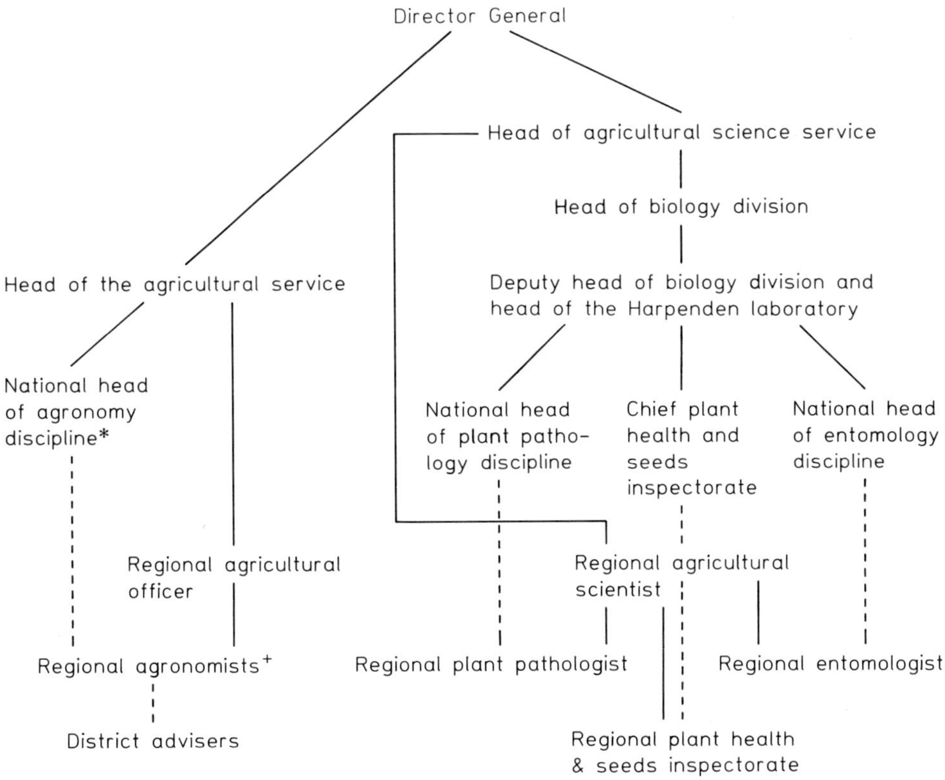

Fig. 3. The management structure of main advisers concerned with crop protection. —— Line management. - - - Functional management. * Refers to the functional management of the Regional Agronomists (arable) only. + Regional Agronomists here include specialists in both agricultural and horticultural crops

Crop protection is not a single entity and is managed in its component disciplines (Fig. 3). For entomology, plant pathology and agronomy management is two-fold, through the national Head of Discipline where matters of technical policy are determined and through the Head of Region, the Regional Agricultural Scientist or the Regional Agricultural Officer where financial matters or co-ordination of work between disciplines or services are the major concern. Each Regional department has an allocation of money for travel, subsistence, R & D and promotional work and must work within this budget. Forward planning is done to provide for major capital expenditure should it be considered necessary. One of the functions of the Regional Management Board in each Region is the overall financial management of the Region. The Board also broadly determines priorities and allocates money to each of the Heads of Services within the Region. Heads of Services can then allocate money to Heads of Departments.

3.1.9 Grade Structure and Promotion Within ADAS

The present grade structure has evolved over the years and is a mixture of two different systems. Initially, there were advisory grades for the advisory service and scientific grades for the research service and other government run scientif-

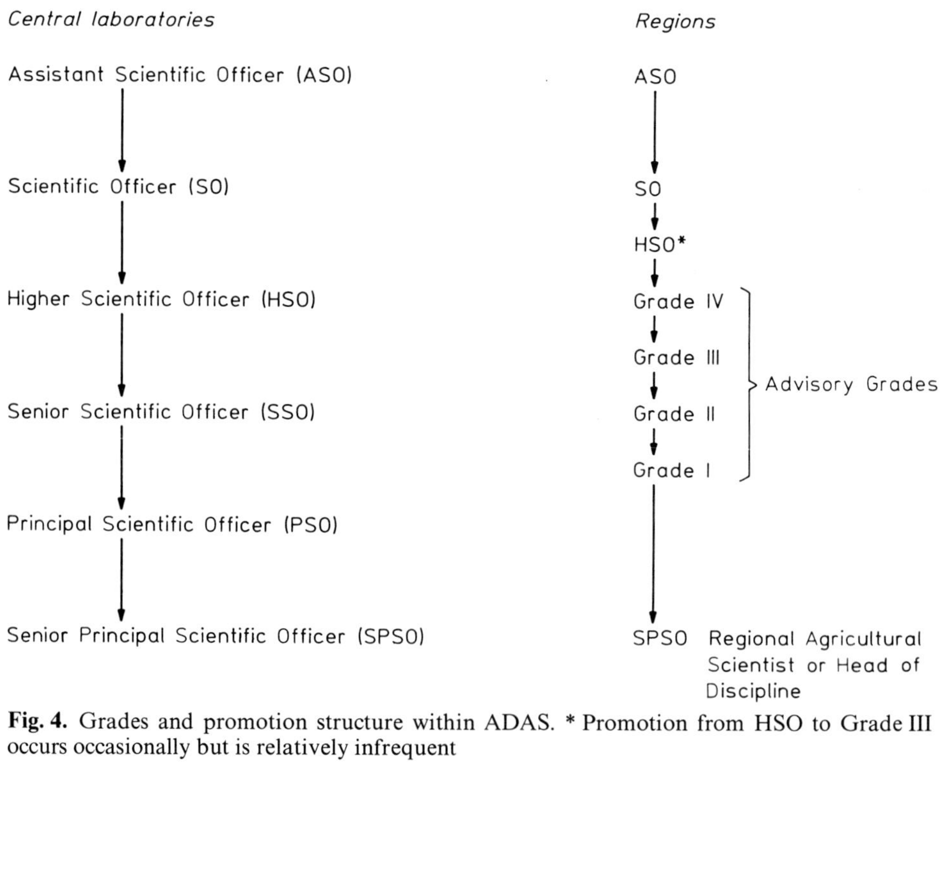

Fig. 4. Grades and promotion structure within ADAS. * Promotion from HSO to Grade III occurs occasionally but is relatively infrequent

ic services. ADAS has a mixture of both grade systems: in the central laboratories all the staff with one exception are scientific grades and the Regional Agricultural Scientists and the Heads of Disciplines are graded in the scientific system as Senior Principal Scientific Officers (SPSO) whilst the Regional Specialists are Advisory Grade I. There are three other advisory grades, II, III and IV, the latter being a training grade. The support staff in the Regions are in the scientific grades of Higher Scientific Officer (HSO), Scientific Officer (SO) and Assistant Scientific Officer (ASO) (Fig. 4). It is possible to be promoted through all these grades from ASO to SPSO but this happens very infrequently. To obtain entry to ADAS advisory grades in the crop protection disciplines requires a good Honours Degree and in practice many new entrants have a Doctorate of Philosophy. To obtain entry into the ASO grade requires a minimum of four subjects at Ordinary Level of the General Certificate of Education (usually obtained at the age of 16 years), but in practice some new entrants into this grade have a degree. Promotion from one grade to the next is done by a system of annual reporting through heads of departments and those advisers with promotion markings are interviewed when vacancies occur. Only rarely are appointments made into advisory grades III and above from outside of ADAS. A promotion within Regions of ADAS almost invariably means a move and many senior advisers have worked in four or five different ADAS laboratories.

3.1.10 Crop Protection in ADAS, Its Strengths and Weaknesses

Agriculture in England and Wales is a highly developed industry requiring a highly specialized advisory service. ADAS has evolved to its present position by meeting the continuous demands of the farming industry. Undoubtedly one of the major strengths of ADAS in general and the crop protection disciplines in particular is the ability to adapt to change and to respond quickly to answer questions posed by new situations. Because of its national organization, this can be done and information can be obtained and put into practice with minimum loss of time. The results of the eyespot resistance survey already mentioned have enabled a more sensible approach to be taken to the control of this pathogen using the fungicidal tools that are available and avoiding possible disasters through lack of disease control. Such a survey could not have been done without a very rapid response to a new situation.

Most farming communities need a reliable source for the rapid identification of pest and disease problems, and also look for an independent opinion when considering the use of pesticides. The service provided by ADAS in examining plant material and soils and the performance of pesticides, rapidly reporting the results of examinations to farmers, provides the industry with such an independent information service. An independent view of this type guards against the indiscriminate use of pesticides and also provides the industry with information on the most effective use of the material available.

Although the direct role of ADAS crop protectionists as advisers diminished over recent years and has been taken up by private consultants or technical representatives there is still a strong ADAS influence on farmers through non-governmental advisers either via the Plant Clinic, which is used by them, or through the written word or by liaison and direct contact.

Strength in depth is perhaps one of the ingredients for the success of any advisory service and the access the ADAS crop protectionist has to laboratory facilities and specialists in other disciplines enables him to provide the farmer with information on most problems that occur.

No advisory system can meet all the demands of an energetic agricultural industry and ADAS is no exception. For instance, since cereal prices have risen and inputs increased to maximize yields, many farmers are using pesticides to a greater extent than ever before. The increased use of pesticides has resulted in a large demand for information on their use. There are not enough district advisers, let alone crop protectionists, to meet such a demand, and the vacuum created has to some extent been filled by private and commercial advisers aided by ADAS technical literature and many group meetings. Literature is no substitute for individual contact, particularly with a generation of farmers used to such contact. It is possible that succeeding generations of farmers will rely less on visits and more on other means of obtaining information, such as TV-based data sources. Decreasing resources, financial and human, are major limitations to the most effective use of crop protection advisers. In such situations it is imperative to use what resources are available to the best possible advantage. The starting point in such strategies is a clear definition of the agricultural situation which needs input, whether in crop protection, marketing or any of the many other aspects ADAS deals with. Once the area for work has been delimited every effort can be made to devise a programme to obtain relevant information. This approach is to be preferred to one of making arbitrary limitations to inputs whilst trying to maintain the same breadth of output.

ADAS does not treat crop protection as a discipline; should it do so? The present arrangement relies heavily on the initial expertise of the general district adviser. Plant Clinics overcome weaknesses to some extent, providing the correct field picture is presented. Can a pathologist, an entomologist and an agronomist sufficiently integrate to give the farmer the package of advice which he needs? Much will depend upon the individuals, but for the system to work it is essential to have first class liaison between disciplines. Such liaison in practice is often the function of the district adviser and the higher the calibre of these advisers, the more likely the system is to be successful. Would ADAS be more effective with specialist Crop Protectionists in each Division? In some areas of high intensity production this may be the case, but the large number of pesticides now available to farmers would mean that such specialists would not be able to handle the quantity of information effectively and maintain an almost equal amount of biological knowledge of all pests, diseases and weeds. Where does crop protection stop? Does it involve nutritional disorders and toxicities, soil structural problems and drainage? If so, the breadth of the subject is so large that the adviser can no longer be a specialist. ADAS district advisers are such generalists and a crop protection generalist as an addition does not therefore seem necessary. What is needed in an advanced agricultural system is a group of crop protection specialists who can advise and support the generalists.

Finally, although computerization is without doubt a great help to all who use the equipment and processes fully, there is a danger that machines become fashionable and symbols of status. If full facilities are not available, i.e. programmes with on the spot expertise for modification, a computer can be a retrogressive step. The correct choice of equipment together with the back-up software can turn a time-consuming machine into a time-saving piece of equipment.

ADAS crop protection, like the industry it serves, has evolved. It must continue to evolve over the years to come, otherwise it will become ineffective. The industry needs and will continue to use an effective advisory service as much as ADAS needs an effective industry.

Note Added in Proof. Since this chapter was written there have been a number of changes in ADAS which affect crop protection. Firstly all ADAS literature und services are no longer free and a scale of charges is operated for the diagnosis of crop problems at all Plant Clinics. Organisational changes have resulted in the heads of discipline post being combined with those of the Regional Agricultural Scientists. Finally the sub centres at Shardlow, Cardiff and Bangor have been closed.

3.2 Advisory Work in Crop Protection in the Federal Republic of Germany (FRG)

K. HANUSS and W. BEICHT

All advisory work is, by its very nature, the tending of help for self-help. It offers solutions from which clients can freely choose, according to their inclination, what capabilities, ideas and aims best suit their problems. It is axiomatic that advice is only given to those wishing to be advised.

A description of the present situation of crop protection advice in the Federal Republic of Germany (FRG) should be preceded by some data on the structure of the FRG and on its agriculture.

The FRG is a democratic federation which comprises 11 states (Länder). Governmental authority is vested in these states, and they are charged with the execution of governmental tasks, unless the basic law of the federation provides otherwise.

Advisory and educational work in the protection of crops and stored products are the primary tasks with which the crop protection services of the states have been entrusted by law. But this does not exclude the possibility of other institutions or private agencies dispensing such advice.

Data on the farming population and agricultural production is given in Chapter 3.2.4.

3.2.1 Organization of Official Crop Protection Advisory Work in the FRG

In view of its significance for the economy in general and for farm production in particular, the protection of crops and stored products in the FRG has been entrusted to several bodies. The states are required by law to provide the advice needed for field crops, pastures, vegetable and ornamental gardens, orchards, vineyards and forests. In addition, commercial crop protection advice is furnished by specialists working for manufacturers of products used in agriculture, especially of pesticides and equipment for their application, and by private consultants. An intermediate position is occupied by advisers working for farmer's associations and similar bodies.

As stated above, the states are entitled to legislate in matters of crop protection, provided the federal government has not seen fit to exercise its right to pass relevant laws. To ensure legal and economic uniformity throughout the area, the federal government has exercised this right by legislating that crop protection services should serve the following aims:

a) The protection of crops from pest and disease.
b) The protection of stored agricultural products from losses caused by parasites.
c) To influence the growth processes of crops by means other than nutrition (e.g. by growth substances), and
d) To prevent any harm that may be caused, especially to man and beast, by the use of pesticides or by other measures taken to protect crops or stored products.

Crop protection services have been instituted in each state to dispense official advice as one of several functions with which such services have been charged. Such advice includes, besides farm advice proper, dissemination of information, instructions and material for the education of the public in matters of crop protection; the maintenance of warning services, and monitoring pests and diseases in crops and stored products; testing pesticides, application equipment and other means of protection of crops and stores; dealing with crop protection problems arising in the export of plants and their products and in commerce in pesticides. Advice is therefore seen to constitute only a part of the activities of the crop protection service. Moreover, the service does not deal with commercial crops only: it also attempts to serve home gardeners and to keep an eye on ornamental plants and on shrubs and trees on non-crop land.

The scope of crop protection work in each of its agencies is a matter of action and reaction: action to fulfil the tasks imposed on it by the state and to carry out the advisory work generally deemed desirable and necessary; reaction with utmost expediency and flexibility to the wishes of farmers and to emergency situations which so frequently arise in pest and disease control. Thus, concepts and directions of work, including the structural set-up, largely depend on regional requirements and on the demands of farmers.

For this and other reasons, it is evident that the crop protection service is not a body uniform in its structure and operation. This is readily seen from the list of principal crops and functions covered by crop protection services in the various states, as presented in Table 1. Besides their official services for agricultural and horticultural crops, four states run separate services for pest and disease control in vineyards, and others in forests. The forest protection services will not be described here in any detail.

Structure. As mentioned above, each state decides how to organize its crop protection services, and these are therefore not structured uniformly.

As a rule, the states are divided into districts, and these in turn into counties (Landkreise). In five of the larger states, organization of the crop protection services follows these divisions, but in three other states their area is adapted to that covered by the local chambers of agriculture. In some states the crop protection services are concentrated in one centre, in others the services have been decentralized and work through regional offices. Advice is also given by subregional, local advisers who are sometimes crop-oriented, e.g. concentrate on sugarbeet or deciduous fruit, etc.

Table 1. The principal crops and functions covered by the crop protection services in the eleven states (Länder) of the Federal Republic of Germany

State	Crops and functions
Bavaria	Winter and summer cereals, maize, hop, potato, sugarbeet, pasture, vegetables, grapevine; inspection of plant material and produce
Baden-Württemberg	Winter cereals, maize, sugarbeet, tobacco, fruit crops, nurseries, grapevine
Hessen	Winter cereals, maize, sugarbeet, fruit crops, vegetables, grapevine
Rhineland-Palatinate	Winter and summer cereals, potato, sugarbeet, pasture, vegetables, tobacco, grapevine
North Rhine-Westphalia	Winter cereals, maize, sugarbeet, vegetables, fruit crops, flower and ornamental crops
Lower Saxony	Winter cereals, potato, sugarbeet, pasture, fruit crops, nurseries
Schleswig-Holstein	Winter cereals, maize, rape, pasture, nurseries
Hamburg	Vegetables, fruit crops, flower and ornamental crops, smallholders' and home gardens; inspection of plant material and produce
Bremen	Vegetables, fruit crops, flower and ornamental crops, smallholders' and home gardens; inspection of plant material and produce
Saarland	Winter cereals, maize, vegetables, flower and ornamental crops; inspection of plant material and produce
Berlin (West)	Smallholders' and home gardens, parks and recreational areas; stored produce, inspection of plant material and produce

Budgeting. The foremost source of budgets for crop protection services is the state. In addition, fees and expenses are paid for certain services or operations carried out by the state, and in some states fees for the operation of the chamber of agriculture are also collected. All these fees may constitute 10–40% of the total budget. Advisory work in vineyards in the field of crop protection is likewise financed primarily by the states; fees for pesticide testing also provide some income. Since advice on crop protection problems in vineyards is provided within the framework of general crop advice, there is no way of defining the amount spent specifically for protection of the crop.

It is hardly possible to estimate the time and financial resources expended for advisory work as part of the general crop protection work of the state. Expenditure evidently fluctuates from year to year, and depends on the season, cultural conditions and the regional structure of the farm economy. Since they carry so many other responsibilities, the central bureaux may be expected to devote less time to the dissemination of information than the advisers attached to regional bureaux. Some more detailed considerations of the time spent in advisory work are included in Chapter 3.2.3.

Facilities and Equipment. To carry out its official functions in the supervision of crops and stored products, in diagnosis, investigations and tests, and in the collection, processing and dissemination of information, the state's advisory bodies have been provided with a variety of facilities and technical equipment. The following deserve special mention:

Diagnosis and Related Investigations. The centres of official crop protection advice have at their disposal laboratories for the determination of harmful organisms, whether plants or animals, and for the diagnosis of infectious and non-parasitic diseases. Moreover, some centres have special equipment for the diagnosis of specific viroses, bacterioses and mycoses, for determination of nematodes, and for analysis of pesticide residues, pesticide application technology (see Figs. 1 and 2), biological pest control, etc.

In some cases the laboratories are organized according to the crops they serve, e.g. for the protection of field crop, vegetables or vineyards, or according to groups of pesticides, such as insecticides, fungicides and disinfectants. Some crop protection centres are also equipped with facilities for breeding animals, with growth chambers for work with pathogens, and with phytotrons and glasshouses for the diagnosis and investigations required by the advisory service.

Equipment for Notification and Warning. Centres and regional bureaux for crop protection are equipped with meteorological instruments, and some record and process data by computer, to issue warnings as required. In addition the records of the German Meteorological Service are utilized for prognoses, particularly to assess chances of infection by *Pseudocercosporella* on winter wheat and for the negative prognosis of *Phytophthora infestans* on potatoes (see Sect. 2.4.2.3, p. 77, on potato blight forecasting). Other devices used for prognosis are pheromone traps to catch on the wing Lepidoptera and Coleoptera harmful to orchards, vineyards, field crops, vegetables and forests. Spore traps are operated to determine sporulation and likelihood of infection by pathogenic fungi, such as the downy mildews of tobacco (*Peronospora tabacina*) and of hop (*Pseudoperonospora humuli*), of apple scab (*Venturia inaequalis*), and of canker (*Nectria galligena*). Incubation calendars and infection tables are also used for this purpose. The distribution and population density of harmful organisms is likewise monitored by light traps, sticky traps, trapping funnels, coloured squares and dishes, and various rodent traps.

Forecasts of scab (*Venturia* spp.) on fruit are made by aid of the usual instruments for measuring temperature, relative humidity and wetness periods, and recently also with the commercially available electronic scab-warning device Biomat. For the prognosis of fireblight (*Erwinia amylovora*), records of the amount of rainfall are kept by self-registering instruments so that danger periods can be calculated by computer. Warnings are communicated by telephone or by leaflets or postcards.

Fig. 1. Testing spray application techniques: is spray being applied uniformly and at the desired rate? Determination of the lateral distribution of spray by nozzles mounted on a boom

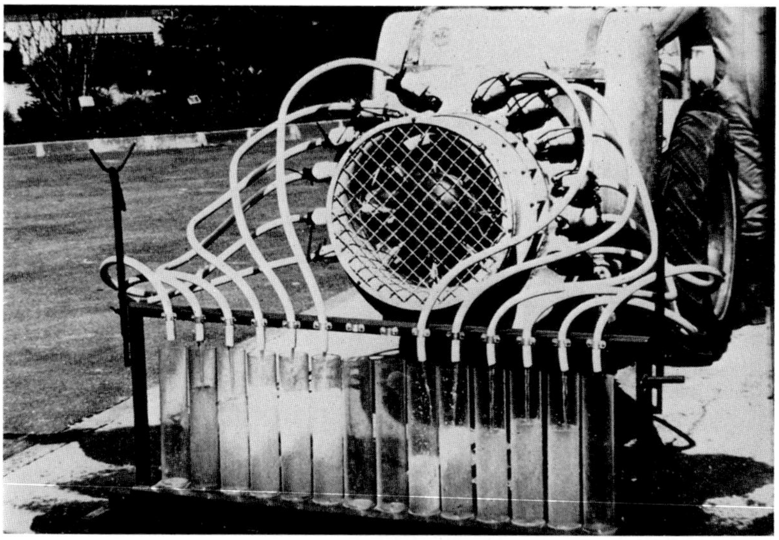

Fig. 2. Determination of the amount of liquid delivered by each nozzle. (See also Fig. 1)

Dissemination of Advice. Information issued by the official advisory bodies is disseminated to their clients by a great variety of methods and pathways (see Fig. 3). Most frequently, advice is given by word of mouth and by written communications to individuals and smaller or larger groups. Electronic media are being used increasingly.

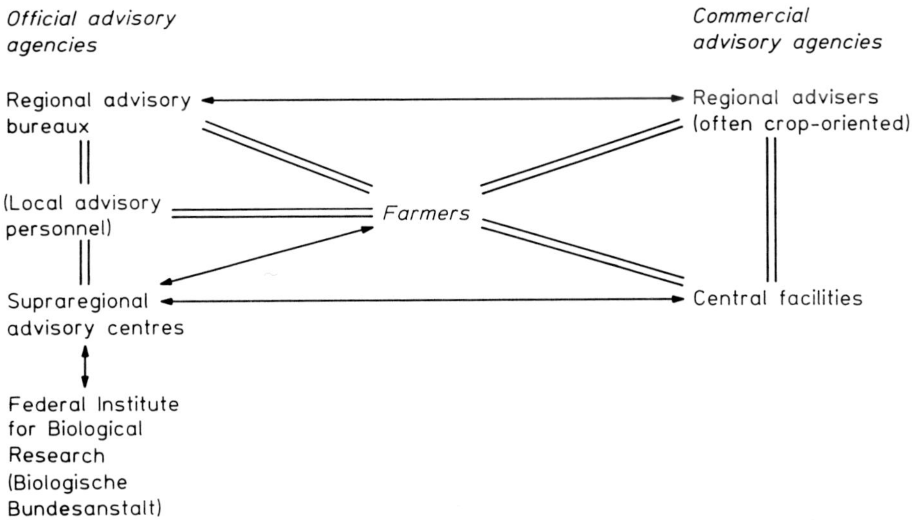

Fig. 3. Pathways of communication in crop protection advice in the Federal Republic of Germany. ═ Preferred pathways of communication. ⟷ Exchange of information in both directions

Side by side with oral advice given directly either in the office or in the field, great importance attaches to communication by telephone, whether by direct calls, or by automated answering services. Field inspections and courses are further means of dissemination of advice.

Occasional or regular announcements over radio or television, in which official advisers take part, afford means of disseminating information to wider circles of the population. In addition, lectures are given and exhibitions held for the benefit of the general public, from the amateur gardener to the specialized grower.

Written information, as mentioned above, is disseminated by the warning service as well as by advisory leaflets on protection of crops and stored products, and by articles in the farming journals and the daily press.

The crop protection services are now increasingly adapting themselves to the use of novel electronic media, foremost among them interactive videotex equipment.

3.2.2 Commercial Advice

Crop protection advice on a commercial basis is given in the FRG primarily by manufacturers of pesticides and of equipment for their application. Advice is also offered by other bodies, such as associations of growers or agricultural cooperatives, or enterprises who employ their own advisers and by manufacturers of farm equipment, as well as by private consultants. But the latter groups are not, as a rule, organized and equipped as well as the advisory agencies of the state or the pesticide-related industry, and the number of such consultants is small; their advice is crop-oriented, including cultural techniques, fertilization, varieties and economic aspects, and is not concentrated on crop protection. Advice is financed by members of cooperatives or clients, the information offered by the official crop protection service and by pesticide manufacturers is utilized for diagnosis and prognosis. Information is disseminated by all the methods indicated above, but predominantly by talking to farmers.

The motivation and aims of such crop protection advice are to further commercial purposes, such as the marketing and sale of pesticides, and they relate principally to the products offered.

Structure and Budgeting. Each of the commercial companies concerned has its own advisory organization. Manufacturers with a broad range of pesticides or application equipment employ specialized advisers, while smaller companies mostly use their sales personnel. The areas in which advisers work may be regional units, and then frequently follow the boundaries of counties, districts or states, or they may be crop-related, e.g. regions in which viticulture predominates.

The expenses of commercial advice are covered by the enterprises concerned. The client is given such advice free of charge.

Facilities and Equipment. For diagnosis and related investigations some of the commercial enterprises have their own research and development departments, but most also utilize the services of research facilities at universities, crop protection bureaux, agricultural research stations, food supervision authorities, etc.

Weather data are primarily obtained from the German Meteorological Service, but some enterprises operate their own meteorological stations. Epidemiological information is derived from personal observations and from the data collected by the official crop protection service.

Dissemination of Advice. Dissemination of advice follows the pathways of advice given by the crop protection authorities. However, written material is often presented in amply illustrated, more costly productions. Oral communications are made by visits, telephone or lectures. Written enquiries are mostly answered by mail, and written information is published in farmers's journals, less frequently in the daily press.

Special importance attaches to supra-regional or regional propaganda activities by means of leaflets, prospectuses, recommendations to growers, technical data sheets, advertisements, professional articles and handbooks.

Printed information is distributed person to person, or indirectly through the channels of wholesale or retail marketing, or by way of professional schools and in advisory or refresher courses. Card indexes, arranged by target groups, are widely used for direct mailing. Advertisements are most commonly placed in farmers' weeklies or monthlies, rarely in the daily press. They are published in rural districts only when of strictly regional importance, or when pest or disease control measures have to be urgently applied on the basis of recent information. Written communications comprise instructions for use and precautions in the use of pesticides. Leaflets are also inserted in journals.

Radio and television are not employed by pesticide manufacturers for advertising large-scale use of pesticides on commercial field and fruit crops, but occasionally for their use for domestic plants and home gardens. Videotex material has not hitherto been used for commercial advice, but is in preparation.

3.2.3 The Adviser

The number and degree of specialization of the adviser depend on the intensity of land use and farm structure in a given region, and on the financial means available. Tasks additional to giving advice often take up a considerable part of the staff's time. The chances for advancing in knowledge and status often differ widely, depending on what the central bureau has to offer.

The Number and Training of Advisers. The government crop protection service consists of civil servants of intermediate, high and superior grades; some comparable functions are taken care of by other employees and technicians. Most of the advisers in intermediate grades have finished technical schools, those in higher grades have completed agricultural high schools. Most of the leading personnel and specialists in the superior grade hold university degrees (in agriculture, horticulture, viticulture, forestry or biology), and are called phytomedical experts (Phytomediziner). This personnel is engaged on complex advisory problems, while those less complex are dealt with by the other advisers and technicians.

In some states acceptance in the crop protection service as phytomedical expert requires preparatory service to qualify for subsequent employment as civil servant in the superior grade. This also applies to applicants with engineering degrees.

Additional voluntary examinations held by the German Phytopathological Society complete the special training in maintaining crop health, a prerequisite for which is a degree in agriculture or biology. In this examination the candidate has to prove comprehensive mastery of crop health problems in general, and of crop protection in particular.

The official crop protection service of the FRG (including West Berlin) comprises about 960 advisory personnel, of which about 23% serve in intermediate grades, 56% in higher and 21% in superior grades. Additional advisers look after health problems of forests and, in some states, of vineyards.

Some 900 crop protection advisers are employed by commercial enterprises in the FRG. This includes about 90 advisers attached to professional and sales departments in regional branches and at the main office. A staff of about 610 works more directly within the framework of the sales organizations and engages chiefly in crop-oriented farm advice. About 200 advisers are specialized crop protectionists.

Out of the total of 900, some 25% hold university degrees, 50% have finished agricultural high schools, and 25% technical or other comparable schools.

Specialization of Advisory Personnel. Advisers of the official crop protection service are in some cases specialized by crops, in others by groups of pathogens or pests. Occasionally crop protection advice is wholly integrated in generalist crop production advice. In some branches of agricultural production advisory work is divided by crops, e.g. in horticulture advice may be given separately to vegetable, fruit, nursery or ornamental crops. Alternatively, specialists may concentrate on advice how to control particular groups of pathogens, such as those causing virus, bacterial and fungal diseases, or of pests, including nematodes, insects, mites, rodents, etc. In most states, specialization by crops is preferred; but in vineyards most advisers specialize by pathogens and pests.

In commercial advisory work, the type and degree of specialization depends largely on the location of the adviser, whether in central installations or at farm level. In the former, staff is specialized mainly according to the objectives to be attained by use of specific pesticides. At farm level, commercial advisers are rarely specialized, though some may preferentially deal with certain crops, such as grapevine or vegetables.

Other commercial or private consultants are not usually specialized and often give advice in crop protection combined with other farm advice.

The Range of Activities of Crop Protection Personnel. The activities demanded from crop protection advisers often range far beyond the sphere of advisory work proper.

Additional tasks given to official crop protection advisers include:

a) Supervision of crops and stored agricultural products to detect pests and diseases.
b) Supervision of the export of plants and plant products, as required by regulations for the protection of crops and stored products, and certifying the health of such crops and products.
c) Dissemination of general information on the protection of crops and stored products, operation of warning services, and advice on the use of biological control agents (Fig. 4).
d) Notification of the occurrence and distribution of pests and diseases.
e) Testing of pesticides and of equipment and methods for the protection of crops and stored products.
f) Supervision of the commerce in pesticides.
g) Investigations and tests resulting from the operations described under (a)–(f), as well as participation in research projects under the aegis of the Federal Ministry of Agriculture and the German Research Association (DFG).
h) Planning and execution of control measures ordered by competent authorities.

Additional tasks are also imposed on crop protection advisers from time to time.

The range of functions to be carried out by crop protection advisers in viticulture frequently includes subjects well beyond pest and disease control, such as research, teaching agricultural schools and colleges, and planning of vineyards and their fertilization.

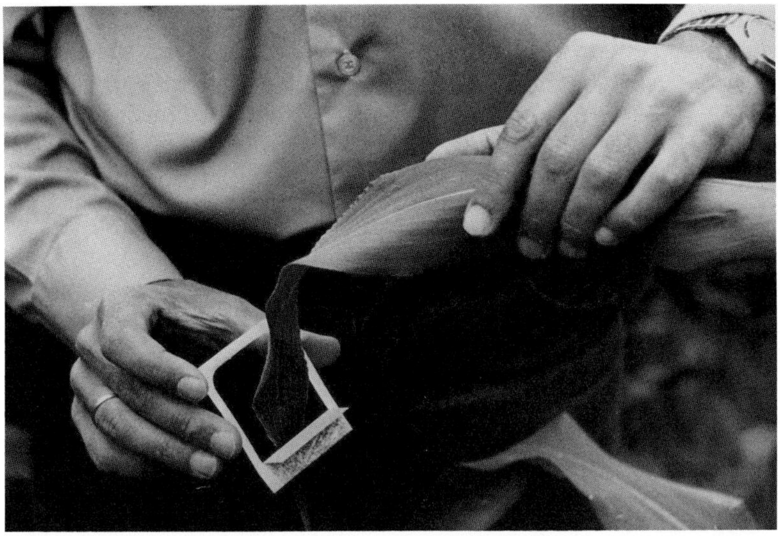

Fig. 4. Advice of biological control is an important part of the crop protection adviser's work. Here he demonstrates how to attach to a maize leaf a carton frame bearing on its lower part about 1500 pupae of the parasitic *Trichogramma* wasp, from which adults are just about to emerge; the flap effects protection from rain. Fifty such frames per hectare can be attached in 20 min, and two such treatments are applied in the course of the season

According to the scope of the enterprises they work for, the tasks of commercial advisers may include crop nutrition, animal nutrition, fodder conservation and sometimes animal hygiene.

Furthering the Adviser's Professional Proficiency. The high standards demanded of the adviser in his work require constant furthering of his professional competence and access to up-to-date sources of information. This is chiefly being taken care of by the centres of crop protection in each state or district, in a comprehensive programme of educational events and in the steady supply of information.

Members of the official crop protection service are offered the following means to this end:

a) Regional exchange and updating of information.
b) Seminaries, conferences and courses.
c) Field demonstrations.
d) Lectures and other means of instruction.

Information is supplied by the spoken and written word, and by audiovisual means.

The frequency of events in such refresher programs varies. Regional meetings for the exchange of information are held once a week to once a month. Field demonstrations obviously depend on seasonal and cultural factors. Comprehensive refresher courses that deal with chosen subjects in greater depth and last 1 – 5 days, are held once or several times a year.

Commercial advisers are offered a varied and often comprehensive program of refresher courses and other events. In addition to constant exchange of information during his current work, the adviser participates in internal and external congresses and courses. Further sources of information are available by contact with research and teaching institutions and with the official crop protection service. This includes utilization of the warning and information services and of the more important literature. Organized refresher courses, lasting up to 25 days, are held two to four times a year.

3.2.4 Targets and Range of Current Advisory Work

The scope of advisory work in the FRG is best illustrated by quoting a few statistics.

The country covers an area of 24.9 million ha. 12 million ha are used for agriculture, including more than 11 million (98%) for field crops and pasture, 120,000 ha (ca. 1%) for horticultural crops, and 100,000 ha for viticulture. Forests cover nearly 7 million ha, or 30% of the total area.

The number of farmsteads totals about 780,000, and this comprises a great variety of farm types in the various states. There are also wide differences in the number of auxiliary and of fully commercial farms.

Targets, Target Groups and Modes of Operation. The aim of official crop protection advice is the promotion of effective, economic, and ecologically correct crop protection, by methods conforming to accepted standards of economics, ecology and toxicology, to keep harmful organisms below their damage thresholds. A basis for attaining this target is the proper exploitation of natural factors that restrict crop damage.

The target groups for such advice include:

a) Growers, manufacturers and operators who perform operations for the protection of crops and of stored agricultural products.
b) Legislators and authorities concerned with crop protection.
c) The public at large, as a target group for current popular explanations.
d) All those giving advice on crop protection matters, e.g. government departments, agricultural associations and cooperatives, employers of rural labour, agricultural trading companies, private consultants, and specialist advisers for all types of crops.

The framework for all advisory work in crop protection is defined by law and official directives. However, there is considerable scope for initiative in the choice of preferential subjects, methods and media. The official adviser is fully and personally responsible for all the actions he takes in fulfilment of his duties.

Directives for his work are in general issued by the higher echelons of the service. They are given by the heads of the state's crop protection service or of the regional bureaux, or by authorized senior advisers, including chairmen of professional committees, or heads of departments dealing with various branches of growing crops, e.g. viticulture. If university institutions have been charged with the performance of advisory activities, their scientists also issue instructions. Directives of a more general nature may also be given after certain professional meetings or conferences.

Commercial advice, especially by manufacturers' representatives, aims at effective crop protection, with the emphasis on control by pesticides. Such advice is essentially similar to that given by the official service, but tends to stress the exclusion of risks and to be specifically crop-oriented.

The target groups of commercial advice are:

a) Agricultural trading companies, cooperatives and employers of rural labour (Lohnunternehmer).
b) Larger agricultural or horticultural farm units and vineyards.
c) Official and inofficial advisers who, in their turn, advise farmers on matters of crop protection.

Commercial, just like official, advice is given strictly within the framework of the law. Advisers receive their directive from the heads of their company's professional departments for pesticides or application equipment. Such directives may be given orally or in writing, by description of the product, and by catalogues listing their objectives and modes of employment.

According to the crop protection law, no statement made in pesticide commerce and advertisement is allowed to deviate from the instructions issued officially for any aspect of the use of the pesticide in question.

Instructions issued by commercial advisers attempt to leave much scope for judgement; within this scope, specific regional advice may be given in agreement with official advisers.

The Range of Crop Protection Advice. Official crop protection advice is given by about 1000 persons. However, since many of these fulfil additional tasks (see Chap. 3.2.3), this does not mean that they constitute 1000 advisory units, if we define such a unit as a person fully devoting his work to such advice. The above advisers, as stated previously, have to cover about 12 million ha and some 780,000 farmsteads.

The area generally served by one adviser is within the boundaries of one county (Landkreis), but specialists often cover larger areas, e.g. a specialist on vegetable pests and diseases may work in several counties of the same district or state. The number of farms served depends on the number and type of farm enterprises in the region: for field crops this number has been stated to range from 420 to as much as 5400, for vegetable and garden crops from a few to 330 such farms per adviser. The frequency with which the same farm is revisited by the adviser differs widely: in some regions only 10–30% of the farms receive recurrent visits, in others as many as 50–70%.

Advisory personnel devotes 10–50% of its time to experimental work chiefly for the solution of specific local problems; about 10% of the time is spent on writing information, instructions or professional publications, but this differs widely with the particular task given the adviser and with his grade: leading advisers may spend up to 30% of their time on such writing.

In vineyards, crop protection advisers are on average responsible for areas of 1500 to 3500 ha. Specialized phytomedical advice for vineyards is generally interregional, and may comprise entire districts in which growing grapevine predominates, or even whole states. Recurrent visits are made to some 60–70% of fully commercial vineyards, and to about 20–40% of those grown as an auxiliary source of income. Here advisers devote about 20% of their time to experiments, 25–30% to courses and lectures and 5% to written information and to publications.

The range of the commercial crop protection adviser's work also varies greatly. It depends on the type of enterprise he works for and the crops concerned. About 10–20% of the time of such advisers is spent on experiments, but the experimental work carried out by development agencies of the companies must also be taken into account. Other activities by these advisers take up approximately the following parts of their time: advice given to individuals 10–20%, to groups 15–20%; tours to collect information 10–15%, lectures 5%, and 2–5% for written advice and for contributing to professional literature.

3.2.5 Evaluation

In conclusion, an attempt will be made to evaluate how well the advice given is received in practice. This can be gauged by the extent to which farmers express their satisfaction. In addition, we shall discuss some suggestions made by advisory bodies to improve crop protection advice and correct shortcomings.

Are the Farmers Satisfied? Official crop protection advice is on the whole accepted by its clients with considerable satisfaction. The crop protection service is regarded as an independent body which relates objectively to all available means of pest and disease control. Farmers' evaluations range from "on the whole satisfactory" to "highly satisfactory". Where reservations were expressed concerning the way farmers are served, they related chiefly to the following aspects:

a) The number of farm units covered is too low.
b) The frequency of visits is not high enough.
c) The area covered by individual advisers is too wide.
d) Limitations in time and funds prevent advice from being given as fully as would be desirable wherever needed, i.e. in fields and stores, and in transit.

The service for pest and disease control in vineyards is similarly rated as "generally good" to "very good".

Commercial advice, especially that given by manufacturers, is also received with much satisfaction, particularly advice on pesticides, and frequent individual advice to large farms.

What Can Be Improved? Room for improvement is generally considered fairly limited, if possibilities of personnel and budgets are regarded realistically. Some proposals and requests are listed here.

In official crop protection work, the size of districts for which individual advisers are responsible should be reduced. The number of farm units visited should be increased, especially in regions with numerous small farms. Growers of intensive crops require more advice from specialized, interregional advisers on protection of fruit, vegetable and ornamental crops, and of crops such as hops and tobacco.

In many cases the work load of the crop protection adviser could be lightened by help provided for his experimental work and by having simple crop protection advice given by technicians.

It has variously been requested that advice by telephone should be augmented, whether by automatic reply services in crop protection centres or by announcements available through the postal services.

As a new information medium, videotex (BTX) is now being introduced in the FRG. Official as well as commercial advice will employ this medium to create easy and rapid ways for clients to acquire information. In a future phase, it is hoped to facilitate question and reply services by computer and videotex and thus provide prognoses and help in decision-making in crop protection matters.

Realization of the concepts of integrated crop protection involves additional tasks and requires suitable provision. The crop protection administration has to be in a position to make the principles of integrated protection available to as many growers as possible. The following steps are essential for this purpose:

a) Increase of personnel for information, advice, education, forecasting, and applied research for the development of all the equipment needed for integrated crop protection.
b) Training of advisers in ecology, cybernetics and data processing.
c) Installation in crop protection centres of videotex and computer equipment to ensure rapid exchange of information, and their linkage with data processing facilities and information retrieval systems.
d) Development of application programmes and their current supervision.
e) Phytosanitary supervision of all rural areas, including those neither used nor reserved for agriculture, to observe their pest fauna and to conserve them as shelter for beneficial organisms.

The following ways of improving commercial advice are viewed as realistic, especially by manufacturers of pesticides and application equipment: integration of crop protection advice in comprehensive schemes for the optimization of crop production; development of electronic data processing and videotex equipment in aid of advisory work; the latter are expected to intensify contacts with clients, and to widen the dissemination of information on chemicals and equipment used in agriculture among their users.

3.3 Advice on Pest and Disease Control in Israel

J. GOLAN and J. PALTI

3.3.1 Background

The principal aspects of crop pest and disease work in Israel are covered as follows:

a) Legislative and regulatory aspects are handled by the Department of Plant Protection of the Ministry of Agriculture, on the site of the Agricultural Research Organization, Bet Dagan. The Department deals chiefly with quarantine, all licensing operations (pesticides and their application), environmental considerations and supervision of the health of propagating material; it is also in charge of certain regional pest control operations, e.g. control of rodents or bats, and carries out field research in vertebrate control and some other fields.
b) Research in all aspects of pest and disease control is carried out at the Ministry's Agricultural Research Organization, with its centre at Bet Dagan, near Tel Aviv, and branch stations in the north and the south of the country. The Plant Protection Institute of this organization has departments for phytophatology, entomology, virology, nematology, pesticide chemistry, toxicology and the study of harmful vertebrates. Research on pesticide application is carried out at the organization's Farm Engineering Institute. Parallel research in the biology and control of parasites is in progress at The Hebrew University's Faculty of Agriculture and in botany and zoology departments of three other universities.
c) Advisory work in crop protection at the Ministry of Agriculture is carried out by various branches of the Extension Service: The Division of Plant Protection does this work on all crops but citrus, the Division of Citriculture on citrus, and the Division of Farm Mechanization gives advice on application techniques. The special arrangements made for the citrus crops are due to the fact that, when the Extension Service began to operate in the early 1960s, extension work in citriculture as a whole centered to a considerable extent on problems of biological pest control; this necessitated visits to groves at a frequency the regional plant protection adviser could not provide, and on the other hand, the work of the adviser in citriculture would have been emptied of much of its content if he had not dealt with control aspects.

Pesticide distributors in Israel dispose of a field staff three to four times as large as the government extension staff giving pest and disease control advice.

Relations are usually quite amiable, and the fact that the government staff is involved to some extent in the pesticide licensing process helps to keep them so: the distributors' staff is interested in having the Ministry's advisers supervise the tests that they carry out for the purpose of obtaining licences. The government extension service, in cooperation with the heads of the staff of pesticide distributors, organize annual training and refresher meetings for the distributors' field personnel.

3.3.2 Availability of Auxiliary Services

Three types of auxiliary service aid the advisory work: plant clinics, a meteorological network, and monitoring (scouting) networks in some of the major crops.

Plant Clinics, each with a staff of one or two, are run by the Agricultural Research Organization at its centre (Bet Dagan) and at its station in the Beersheva region, as well as by the Ministry's Department of Plant Protection. The clinics deal primarily with material suspected of infection by fungi and bacteria, and their work includes occasional pathogenicity tests with isolates from such material. In addition, the staff of these clinics submits samples suspected of infection by virus or infestation by nematodes, insects, mites or vertebrates to the appropriate research departments. The clinics maintain card indices with records of disease occurrences, specifying details of occurrence such as locality, soil, cultural aspects (irrigation), pesticidal treatments applied, and incidence of disease.

An important widening of activities has recently taken place at the Gilat (Beersheva) clinic, which now offers farmers four services against a slight fee:

a) The mapping of fields suspected of infection by *Verticillium:* farmers are given a map of locations at which soil samples are to be taken in a given field, and are instructed how to take samples and send them to the clinic. They are informed within a few days how infected the field is, and whether soil disinfection is advisable or not. Since potatoes and cotton, both highly susceptible to the disease, are two main crops in this region, the service is getting very popular.
b) Examination of consignments of potato set tubers grown in Israel or abroad, to ascertain that they are free from diseases, especially *Verticillium.*
c) Testing of field and orchard crops for nematodes.
d) Examination of avocado propagation material to make sure it is not affected by *Phytophthora cinnamomi.*

Meteorological Service. A network of meteorological stations is run by the Ministry of Transport. These include about 20 agrometeorological stations in which, in addition to standard humidity and temperature levels, records are also taken of parameters of special interest to agriculture, such as duration of dew cover, soil temperatures, etc. These stations could readily be used for disease and pest forecasting work. However, under the semi-arid conditions of Israel, forecasting of diseases such as downy mildews and late blight (*Phytophthora infestans*) on the basis of meteorological data alone, has not been successful (Palti and Rotem 1973). Thus predictions are so far limited to projecting effects of weekly or monthly mean temperatures on subsequent appearance of certain pests and diseases (see table in Sect. 2.4.2.1 on pre-season criteria of parasite appearance), and in drawing-up calendars of periods free from certain diseases. In Israel's semi-arid climate, where many crops are irrigated for much of the year, no systems have so far been found to utilize meteorological data as basis for regional forecasts of actual disease or pest appearance within short spans of time (days or weeks).

Monitoring or Field Scouting. On two major crops, cotton and citrus, pest monitoring networks are in operation and are financed by the semi-government Marketing Boards for each of these crops. Cotton fields are scouted about once a week, pest populations are recorded in relation to well-defined treatment thresholds, and the operation is aided by pheromone trapping of moths. Regional chief scouts supervise this work and results are discussed at weekly meetings in which the regional crop protection adviser frequently participates. Citrus groves are scouted at intervals which depend on the season, and records are taken of the populations of pests and beneficial insects or mites. The California red scale (*Aonidiella aurantii*) is monitored by pheromone traps. Advisers of the citriculture division work in close contact with these scouts.

Fig. 1. Traps consisting of yellow sticky boards are widely used in Israel to monitor the flight of the olive fly (*Dacus oleae*) and under certain conditions also serve to control the pest

In a few regions pest and disease scouting on crops of special regional importance has been organized by local growers' committees or regional authorities, usually with the active cooperation of the crop protection advisory staff. In particular, in the Jordan and Beisan Valleys, seasonal scouting has been instituted on a number of crops: on olives, the olive fly (*Dacus oleae*) is trapped by yellow sticky boards (Hameiri 1973) (Fig. 1), on vines diseases and pests are monitored by frequent visits, and on wheat, rusts and *Septoria* blotch are similarly observed and desirability of fungicidal treatments is determined accordingly (Dinoor and Shtienberg 1983) (see Sect. 2.4.6.3). In the Upper Galilee, a monitoring service has been organized for apple orchards. It may be noted that development of monitoring on crops other than cotton or citrus is largely limited to regions in which collective settlements (kibbutzim) constitute an important element in the agricultural community, since such settlements are most cooperative in such enterprises.

3.3.3 The Crop Protection Adviser in the Extension Service's Division of Crop Protection

Organization. The division is headed by a director and three senior advisers with headquarters at the Ministry of Agriculture, Tel Aviv, aided by a coordinator for group activities and publications. In each of the 10 regional bureaux of the Extension Service there are one to three regional plant protection advisers (Table 1). The staff of the division totals 23.

Almost all the advisers are academically trained. They deal only with extension work, including pest, disease and weed control, and not with administrative tasks, such as law enforcement, valuations for tax and other purposes, etc.

The Senior Advisers. These advisers are specialized on groups of crops, one each dealing with field and vegetable crops, fruit crops (except citrus), and flower and ornamental crops. The tasks of this senior staff are manifold, and may be described as follows:

a) *Field Work.* Supervision of the work of regional advisers; supplementing the latters' work in conjunction with the appropriate crop specialists (see below); participation in research, to develop treatment thresholds and clarify details of pesticide use under special cultural conditions, and to develop integrated control.
b) *Recommendations.* Each senior adviser heads a committee that formulates annual recommendations for pest, disease and weed control in the crops in his charge. Research staff, growers and crop advisers are represented on these committees. The recommendations refer, of course, only to duly licensed pesticides, but take into account not only their efficacy against parasites and phytocidal properties, but also price, convenience in use, compatibility, toxicity, environmental factors, etc.

Table 1. The range of crops served by regional crop protection advisers

Region (from north to south)	Crops most prevalent in the region	Crops served by the crop protection adviser, and additional tasks[a]
1. Upper Galilee	Deciduous fruit	1. Fruit crops only 2. Fruits crops, cotton, vegetables
2. Western Galilee	Subtropical fruit	1. All fruit, cotton, vegetables; also acts as country-wide adviser for subtropical fruit 2. Cotton and tomatoes for processing
3. Jordan and Beisan Valleys	Cotton, olives, wheat	1. Olive, cotton, vegetables; also acts as country-wide adviser for olives 2. Subtropical fruit, vegetables
4. Nazareth	Vegetables, field crops, olives	Vegetables, field crops
5. Northern coastal plain (Hadera)	Cotton, vegetables, flowers, vine	1. All field and vegetable crops; also acts as adviser for cotton in some other regions 2. Flower and fruit crops
6. Yezreel Valley	Cotton, vegetables, cereals	Field and vegetable crops, flowers
7. Central coastal plain (Tel Aviv)	Vegetables, flowers	Vegetables, flowers
8. Southern coastal plain and Jerusalem region	Deciduous fruit, cotton, flowers, vegetables	1. Vegetables, field crops 2. Flowers 3. Cotton
9. Lakhish	Vegetables, flowers, cereals	Vegetables, flowers
10. Negev and Arava	Cereals, vegetables, flowers, fruits	1. Vegetables and field crops 2. Protected vegetables and flowers 3. Flowers and vegetables, subtropical fruit, dates; also acts as country-wide adviser for dates

[a] In regions in which more than one adviser is employed, the figures 1–3 indicate the crops with which each one is chiefly concerned

c) *Periodical Publications.* The above recommendations are published by the senior advisers in annual or bi-annual brochures that also contain instructions on safety in handling pesticides, treatment thresholds for certain pests, prescriptions how to monitor crops and other relevant information (e.g. Melamed and Shoham 1977). Current advice on pest, disease and weed control is given in leaflets which summarize the past week's observations and provide warnings of pest and disease occurrence to be expected according to previous experience and to results of monitoring or trapping of pests. Such leaflets are issued weekly all the year round for field and vegetable crops, but only in their main seasons for fruit and flower crops. The leaflets generally do not specify the pesticides to be used, but refer the grower to the appropriate page of the above-mentioned brochures for detailed prescriptions. However, pesticides newly licensed and offering special advantages may be mentioned in the leaflets. The senior advisers also participate in formulating leaflets to be published by the regional bureaux.

d) *Radio, Television and Video Films.* Seasonal pest control recommendations are broadcast regularly. Colour television and video equipment is available at headquarters and at all regional bureaux; the senior advisers actively participate in the production of suitable video material, the subject so far produced including films on calibration of spray equipment, pest scouting in cotton, the *Phytophthora* disease of avocado, soil disinfection by metham sodium, and the Mediterranean Fruit Fly (*Ceratitis capitata*).

e) *Training.* The senior staff is responsible for the programmes of, and often lecture in, refresher courses for various target groups, including crop advisers on the staff of the advisory service, field and orchard pest monitors, pesticide distributors. They also plan, and frequently participate in, courses at various levels for growers of different groups of crops or in the plant protection part of crop-oriented courses. Participation of the senior advisers in foreign training programmes in developing countries and in international crop protection courses held in Israel for trainees from these countries, must also be mentioned as part of their teaching tasks.

f) *Other Activities.* The allotment of priorities in crop protection research is in part in the hands of commodity boards created for this purpose. These boards, which are mostly crop-oriented, include representatives of research institutions, growers' organizations and pesticide distributors' associations, and are attended by the senior advisers. There is one board for each major crop or group of crops, and as far as crop protection is concerned they are concerned chiefly with pests or diseases special to that crop. To deal with more general problems of crop protection and with multi-host pests, there is a special crop protection board. Participation in the work of all these boards is an important part of the work of the senior advisers and of the director of the division.

The work of the pesticide registration (licensing) committee has been referred to above. The senior crop protection advisers are permanent members of this committee and actively participate, at the request of distributors, in collecting evidence for discussion in the committee. This entails a considerable amount of detailed work, as representatives of pesticide distributors seek the advice of this staff in formulating labels corresponding to the decisions of the committee. But participation in this work helps to keep the adviser up-to-date in all developments in the pesticide field.

Last, but not least, mention must be made of the close contact constantly maintained between the senior advisers and all scientists engaged in crop protection research. This is very definitely of mutual benefit: the scientists value the information from the field, which problems deserve priority in their research proposals, and the feed-back on success achieved in the application of research results; the advisers are given the opportunity of early acquaintance with research results, and sometimes with influencing the direction in which research proceeds. Thus these advisers are members of the committees evaluating research proposals before they are funded, and the director of the department is a member of the research evaluation panel of the Agricultural Research Organization. These close contacts between extension and research result in the practical application of research findings with the least possible delay.

Coordination of Group Activities and Publications. The Division's work comprises a large amount of interregional activities, including all the training activities referred to above, monthly meetings of the Division staff, and participation in the activities of other divisions of the Extension Service. This evidently requires detailed arrangements, which is the task of a coordinator working out of headquarters. Weekly publication of advisory leaflets, and periodical publication of the brochures written by the senior staff, are also seen to by this coordinator, aided by the Division's secretary.

The Regional Advisers. Almost all the regional crop protection advisers are academically trained. Like the senior staff, they are not specialized by crop protection disciplines, i.e. they are not extension phytopathologists, entomologists, weed specialists, etc., but their task is to deal with the entire complex of crop protection problems in a crop. Since a great variety of crops is grown in Israel, and many crops under irrigation or under plastic cover are grown several times a year, few advisers would be able to handle all problems in all crops. Their work is therefore divided by groups of crops. The principal division is between crop protection advisers for field and vegetable crops and for fruit crops, with many interactions with flower and other crops. Some of the regional advisers act additionally outside their region proper as country-wide advisers for crops in which they have acquired special expertise. The flexibility of the choice of crops with which regional advisers deal is best illustrated by Table 1, which reflects the actual division of work arrived at by trial and error.

A point of special importance is the integration of crop protection work with that of the crop specialist. As seen from Table 1, the regional crop protection advisers cannot cover all crops in their region. Field and vegetable crops are actually covered all over the country, but fruit crops are not covered everywhere, and flower crops only in about half the regions. Advice on protection aspects in these crops is given by the crop specialist, guided by the senior crop protection adviser for that crop, working out of headquarters, or by those advisers that have assumed country-wide responsibilities. Thus, in the four northerly regions at the head of Table 1, regional advisers in floriculture are guided by the senior adviser for flower pests and diseases; similarly, in the four southerly regions listed last in the table, advice on the protection of fruit crops is given by the regional advisers in fruit growing, guided by the senior adviser for protection of fruit crops and, in the case of subtropical fruit, by the adviser from the Western Galilee in his capacity as country-wide authority for these crops.

3.3.4 Planning the Work of the Regional Adviser

The work of the regional crop protection adviser is planned by the director of the Division in conjunction with the head of the regional advisory bureau. The latter, in turn, is guided by a board on which government and growers are equally represented and which can allot a certain amount of money to local advisory and research activities. Work planning in this way ensures, within the framework of priorities set by the government's farm policy, that specific local problems are given due weight, but that a certain interregional balance is also achieved between the advisers' activities, e.g. that development or experimental work is not unnecessarily duplicated.

The planning takes into account the variety, area, and economic importance of the crops grown, the magnitude of their plant protection problems, the expertise and work capacity of the crop protection adviser, the professional level of the crop specialists and their acquaintance with crop protection problems, and the extent to which such specialists can be guided by senior plant protection staff.

As regards experimental work in the crop protection adviser's work schedule, an important point is the availability of local research facilities: where experiment stations or experimental farms are situated in the region, the amount of adaptive research required from the adviser may be less, but on the other hand, his participation in the station's research is easier and attractive. On average, the regional adviser allocates 30–50% of his time to experimentation, 20–40% to direct farm advice in the field or in his office, and about the same to visits jointly with crop extension specialists. All advisers wish to carry out some experimentation, since exclusively advisory work is repetitive and experiments provide interest and stimulation. But the amount of time advisers in the ten regions listed in Table 1 actually devote to experimentation varies greatly. It ranges from active participation in comprehensive research projects, e.g. in pest biology and control in subtropical fruit and cotton, to the execution of simple pesticide tests.

Numerous field tests are carried out in growers' crops, and in cooperation with the staff of pesticide distributors.

Regional advisers are not required to report on each farm visit. They communicate to headquarters weekly summaries of observations and problems, and these are included in the weekly leaflets mentioned above.

3.3.5 Evaluation and Prospective Improvements

As mentioned above, the present organization of crop protection advisory work has been set up in the early 1960's. The agriculture of Israel has since undergone wide changes, in which the Extension Service has played an important role. These changes include, on the positive side, a large expansion of areas under subtropical fruit (especially avocado, mango, litchi, actinidia) and of vegetable and flower crops protected by plastic foil; the wide use of trickle irrigation; and the rise in the educational level of farmers. On the other hand, there has been a certain regression in citriculture, due chiefly to competition in export markets, and also some reduction in the number of farmers, as many of the collective settlements have developed industrial branches in preference to agriculture.

Some Recent Developments in Crop Protection. In the sphere of pest and disease control, some of the more important events have been: (a) the introduction of soil disinfection by use of solar energy (solarization) (Katan 1980). In a farm economy based on rapid and continuous succession of largely irrigated crops, with little crop rotation, the danger of soil-borne diseases is so prominent that this and other methods of soil disinfection have made valuable contributions, including the control of broomrape (*Orobanche* spp.). (b) Breeding and introduction of resistant cultivars of numerous crops, including wheat, groundnuts, tomatoes, potatoes, vetches, chickpeas, cucumbers, etc. (c) Better integration of cultural and chemical control: where trickling has replaced overhead sprinkling, marked changes have taken place in crop health, including the lessening of injury from salinity and frost (Palti and Shoham 1983), and in many other cases irrigation practices have been developed that reduce the chances of pest and disease attack (Fig. 2). (d) Integrated biological and chemical control of insect pests and spider mites in cotton, cucurbits and avocado, but some setbacks in such control in citrus. (e) Appearance and spread of diseases new to Israel, before all *Phytophthora cinnamomi* on avocado but also *Gaeumannomyces graminis* on wheat, and bacterial diseases such as *Pseudomonas corrugata* on tomato and *P. lachrymans* on cucurbits (Fig. 3).

Specialization and Advancement of Regional Advisers. The operation of the plant protection advisory service described above must be evaluated against the background of these developments. In general, the advisers have been able to adapt to changes reasonably well, and to bring results of local research to the farmer without delay. But many regional advisers fail to keep up with developments abroad, in part because their knowledge of English is inadequate: they do not read even the more important reviews and generally rely on the staff at headquarters to keep them informed.

Has the early decision to have regional advisers specialize on crops rather than by disciplines been wise? In retrospect, it can probably be said that it was right at the time it was taken, because it has facilitated maximum flexibility in the adviser's work. It is easier for an all-round plant protectionist to switch from one crop to another than for a specialized adviser to switch from phytopathology to entomology. The proximity of research facilities and supporting services in so small a country has made it possible for the adviser to call in research specialists and other help where needed. Moreover, specialization by crops was an asset in dealing with health problems against the background of a farming community with little agricultural heritage and in view of the rapid introduction of novel cultural practices.

However, about 20 years after the present system was instituted, the question must be asked whether it is still optimal for crop protection advice at the pres-

Fig. 2. Use of cultural means to check crop disease: covering the soil of strawberry beds with plastic sheeting greatly reduces the incidence of soil-borne fruit pathogens such a *Rhizopus stolonifer*, which causes watery soft rot. (Photo by courtesy of A. Abdul-Razik)

ent juncture of development. The professional level of many growers on collective as well as family farms has improved greatly, and extension aids (audio-visual material, television, radio) are universally available. A change has also taken place in the level of information and competence of the crop specialist advisers. While in the past many crop specialists were experienced farmers without formal training, this is no longer true. Some specialists have studied while doing part-time advisory work, and all the younger personnel is academically trained. They are capable of handling much of the plant protection work in their crops and like to do so.

Fig. 3. Cucumber leaf affected by angular leaf spot (*Pseudomonas lachrymans*). The disease was unknown in Israel until culture of melons and cucurbits under plastic tunnels was introduced, but is now widespread on crops grown in tunnels as well as in open fields

In order to keep abreast of specialized growers and generalist crop advisers, the crop protection adviser must deepen his knowledge. This tends to move him in the direction of specialization by discipline, to become an extension phytopathologist, entomologist, etc. Another, no less important, reason to move in that direction has been referred to in the chapter on the adviser's functions: general crop protection work on one group of crops in a limited region, if carried out continuously for a fairly long period, has been found in Israel to turn to some extent into routine work, and then fails to afford the regional adviser enough of a challenge. On the other hand, the advance in integrated pest management, especially in cotton, avocado and citrus, demands an overall view of pest control problems which more specialized research or advisory staff rarely provides. This is so important that it deserves to be considered very carefully before the adviser is permitted to specialize.

In order to reduce the routine element of his work and provide him with a challenge, without losing any of the above advantages of his generalized knowledge, a beginning has been made in entrusting each regional adviser with countrywide responsibilities for one crop or group of crops with which he has to deal in his own region: in addition to such advisers for cotton, subtropical fruit, and dates, as mentioned in Table 1, various regional advisers are being given the task to deal more comprehensively with certain crops, e.g. leaf vegetables, hothouse crops, processing tomatoes, etc. This includes current reading of literature on crop protection problems of these crops, presenting from time to time summaries of such literature, participating as representative of the department in committees on research planning and formulation of recommendations for these crops, and in initiating and carrying out appropriate field tests. One day a week is allocated for this extra-regional work, to allow for visits to fields, research stations, libraries, etc.

Subjects to Be Integrated in the Activities of the Division of Crop Protection of the Advisory Service. A number of subjects not so far included in the Division's activities should, on the basis of experience gained, be integrated in its operations. The most obvious is the case of citrus: crop protection work on this crop should in future be included in the general scheme of work of the regional or bi-regional plant protection specialist. There is no longer any justification for keeping work in this crop separate from all others; most aspects of integrated pest control in citrus can now be handled by the citrus crop specialists, and the crop protection adviser can back them up.

An area that requires fuller coverage is that of pesticide application. Extension work in this area should definitely be transferred from the Farm Mechanization Division, where it receives somewhat cursory treatment, to the Division of Plant Protection. A senior adviser for pesticide application should work at headquarters in close conjunction with the other senior staff and with a scope of work similar to that of this staff.

The economics of plant protection operations, including the definition of damage and treatment thresholds, has not so far received adequate attention in the work of crop protection advisers. With the constant rise in the cost of control by pesticides relative to total crop expenditure in many crops, this aspect merits more thought and input in the training of crop protection advisers and in their work.

References

Dinoor A, Shtienberg U (1983) Optimization of chemical control and yield loss assessment of leaf rust and Septoria tritici in wheat. Abstr Proc 4th Int Congr Plant Pathol, p 214

Hameiri Y (1973) Colour traps for the olive fly. Isr Minist Agric, Ext Serv, 22 pp. (Hebrew, with English summary)

Katan J (1980) Solar pasteurization of soils for disease control: status and prospects. Plant Dis 64:450–454

Melamed Y, Shoham Ch (1977) Insect scouting in cotton fields. Isr Min Agric, Ext Serv, 17 pp

Palti J, Rotem J (1973) Epidemiological limitations to the forecasting of downy mildews and late blight in Israel. Phytoparasitica 1:119–126

Palti J, Shoham Ch (1983) Trickle irrigation and crop disease management. Plant Dis 67:703–705

3.4 Plant Protection Advisory Work in the University of California Cooperative Extension Office

I. J. THOMASON and NICK C. TOSCANO

California is the leading agricultural and urban state in the United States of America. More than 200 commercial crops are grown on 3.94 million irrigated ha with an additional non-irrigated 11.8 million ha devoted to field crops or grazing. The state's gross annual agricultural production was $ 14.6 billion in 1982 (Table 1). The annual loss due to pest damage is estimated at $ 2.2 billion.

The University of California (UC) serves the state's agricultural sector by developing research on such critical areas as cultivar improvement; harvest and post-harvest practices; plant physiology and ecology; plant nutrition; soil fertility; irrigation; animal health, nutrition and reproduction; and pest management practices for insects, pathogens, nematodes, and weeds. Cooperative Extension (CE) is a statewide program of the University of California Division of Agricultural Sciences which serves as a link between University-based research and the state's agricultural producers.

Table 1. Gross value of plant and animal production in California 1982 crop year

	Dollar value in thousands
Vegetable crops	$ 2,641,340
Seed and nursery crops	1,125,422
Cereal grains	934,653
Alfalfa, irrigated pasture, and wild and grain hay	803,394
Cotton lint and seed	1,110,291
Dry beans, safflower and sugar beets	257,649
Deciduous fruits and nuts	1,199,432
Citrus, avocados, and olives	955,053
Grapes	1,316,117
Strawberries	246,084
Total crop value	$ 10,589,436
Cattle and calves	1,205,973
Milk and milk products	1,905,888
Chickens, turkeys, and eggs	813,390
Hogs, sheep, and wool	108,688
Total animal value	$ 4,033,940
Total plant and animal production	$ 14,623,375

3.4.1 Cooperative Extension Work in California

The mission of Cooperative Extension in California is to identify and solve problems for people located on farms and in communities beyond the nine campuses of the University of California system by extending new knowledge created through research. Where an adequate research base is not available, Cooperative Extension brings needs and information from individuals and communities back to campuses for further exploration and research. This is accomplished through the efforts of more than 150 campus-based academic specialists and 300 farm, marine, forest, home, and youth advisors located in 47 county-based offices. The advisors identify local needs and opportunities for research application and use their own academic expertise to conduct adaptive research and develop educational programs to bring about the use of new information and technology.

The major responsibilities of the crop protection advisor are to: (1) establish and maintain contacts with the agricultural community; (2) establish and identify pest management needs of the agricultural community in his/her county; (3) conduct research cooperatively with other farm advisors, campus-based Cooperative Extension pest management specialists and Experiment Station personnel; (4) disseminate all useful and timely information to the local community by the use of tours, meetings, media resources, and personal contacts; and (5) stay abreast of current research and technology on integrated pest management practices by attending training/educational sessions and taking sabbatical leave when necessary and possible.

To give a practical example of how this works, let us assume that an Entomology Farm Adviser, who usually holds a Master's Degree in Entomology, observes white fly (*Bemisia* sp.) causing serious damage to melon or lettuce crops, and possibly transmitting viruses. He sets up a project to monitor white fly populations, populations dynamics, and overwintering hosts. He then calls in plant pathology and entomology specialists from a university campus, to develop liaison between experiment station research scientists and those involved in field activity.

The role of the campus-based specialists is thus to serve as resource persons for county-based staff by maintaining expertise in research and its potential application. They develop solutions to locally identified needs, translate local definitions of need into technical research language and ensure that field needs will be considered among Agricultural Experiment Station research priorities.

The specialists are assigned to one of the eight program areas of Cooperative Extension, and are based primarily in research and teaching departments at Berkeley, Davis, Riverside, or the Kearney Agricultural Research and Extension Center. The eight CE programs are: (1) Agricultural Economics and Community Resource Development; (2) Agronomy, Range, and Vegetable Crops; (3) Animal, Avian, Veterinary, Forestry, Wildlife, and Marine Sciences; (4) Horticulture; (5) Pest Management; (6) Soils, Water, and Agricultural Engineering; (7) 4-H Youth; and (8) Food, Nutrition, Family, and Consumer Sciences.

3.4.2 The Pest Management Program

The Pest Management Program area of the University of California Cooperative Extension (UCCE) seeks to develop and bring about the use of environmentally and economically sound pest control practices that will mitigate food and fiber losses and reduce nuisance and health concerns. With the growing public concern for safe use of pest control chemicals and the rising cost of chemical treatment, high priority is given to development and application of integrated pest management (IPM) principles.

The Pest Management Program encompasses the following disciplines:

1. Entomology – management of insects and related arthropod pests of plant and animals. This includes work with beneficial arthropods, such as honeybees and other pollinators, parasites, and predators that feed on destructive insects.
2. Plant Pathology and Nematology – management of plant pathogens and nematode infestations.
3. Weed Science – management of undesirable vegetation.
4. Pesticide Management – coordination of pesticide-related programs.

Additional special programs include vertebrate management, environmental toxicology, and mosquito control.

Entomology, plant pathology, nematology, and weed science emphasize development and adoption of IPM systems for control of pest insects and related arthropods, plant diseases, and nematodes in agricultural production and urban areas, and for control of undesirable vegetation in agricultural, wildland, and urban areas. In the process, staff in these disciplines work closely with a special cross-disciplinary team of integrated pest management (IPM) specialists (Flint 1981) (Fig. 1). Programs emphasize development of nonchemical pest control methods such as cultivar selection and cultural modification, as well as evaluation of chemicals and chemical application techniques. Chemical evaluation focuses on determining conditions for safe use, residue levels, persistence, environmental risks, and efficacy. Since California growers are producers of many specialty ornamental, vegetable, tree, and vine crops, they tend to receive special attention from advisors and specialists in the Pest Management Program.

With 36 specialists and 15 advisors, the Pest Management Program serves a broad spectrum of clientele. Entomology, plant pathology, nematology, and weed science groups direct their programs toward the 82,000 owners and operators of farms in California. Of these, 66,000 are individuals or individual families, another 11,000 are partnerships, and the remainder are corporate entities, including incorporated family farms. The specialists in the pesticide management section direct their efforts toward these farm owners and operators, the general public, and the estimated 220,000 farm laborers employed in the state.

Weed science specialists and advisors also direct their program toward the managers of public and private forest, range, wildland, and landscaped grounds. These individuals, numbering more than 5000, exercise control over approx. 4,200,000 ha of forests, 4,200,000 ha of rangeland, and 8,400,000 ha of

Fig. 1. Area IPM Specialist Philip Philips (*left*) discusses monitoring techniques for citrus thrips with growers

wildlands (including irrigation and drainage reservoirs, canals, and rights-of-way).

The Pest Management Program as a whole also serves privately employed professionals in the pest control field, including licensed pest control advisers (about 5800 in the state) and pesticide applicators (about 10,000), as well as representatives of public agencies charged with the regulation of pesticides. These groups are regarded as secondary audiences in as much as their participation in programs is sought because of their impact on agricultural production.

The Pest Management Program ultimately serves all California residents who can utilize pest management information to improve or enrich some aspect of their lives (potentially all 17 million adult residents of the state). Included among these are several million non-commercial agriculturists (gardeners and livestock owners), who raise crops or animals for their own enjoyment or consumption, and thus have a special interest in pest control matters.

3.4.3 Planning and Conducting Pest Management Programs

The Process of Program Planning. In order to best serve clientele needs, Cooperative Extension has developed a coordinated process of program planning; once priorities have been established, resources are directed toward developing programs to meet those needs. Priorities are set largely by workgroups made up of both discipline- and commodity-oriented specialists, advisors, and University researchers. By bringing together these key individuals, with input from clientele groups, the workgroup can:

1. Assess clientele needs/problems.
2. Identify applicable research or research gaps.
3. Establish priorities.
4. Identify, design, and carry out major projects for the development of:
   - Relevant research (if necessary)
   - Application of research
   - Adoption strategies
   - Dissemination of information.
5. Evaluate impact of major projects.

This process is probably best illustrated by the following model (Fig. 2), which was developed by Associate Director-Programs James M. Meyers (Meyers 1982).

In planning and conducting its programs, the Pest Management Program has identified four general problem areas: (1) providing education programs, (2) monitoring and evaluating pest problems, (3) methods of management, and (4) environmental and economic impact of pest management. Each of these broad categories is described below:

Educational Programs. Providing educational programs is an essential ingredient in accomplishing the goals of the Pest Management Program. Growers, pest control advisors, and the public sector must be aware of recommended management strategies in order for the program to be effective. Nearly 40% of current programmatic resources are devoted to these activities.

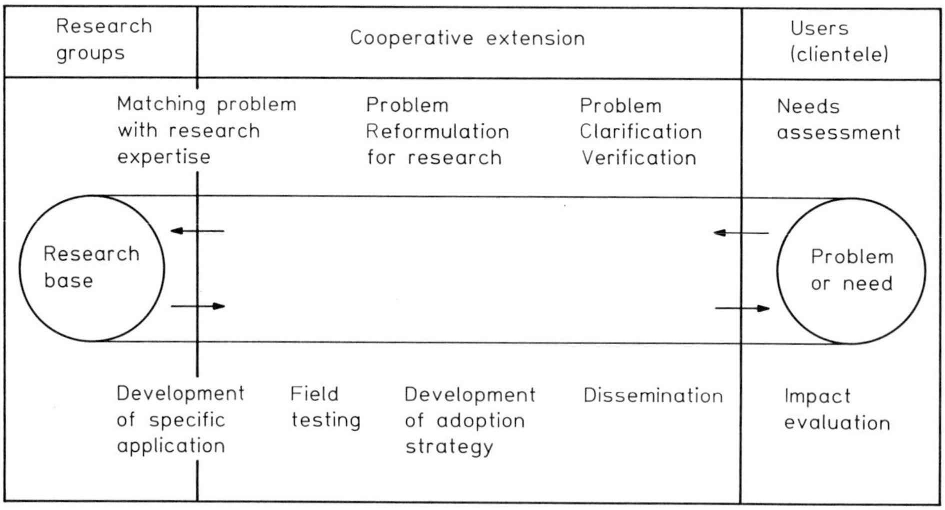

Fig. 2. Model of program planning (Meyers 1982)

An increasingly important method of transferring information is the use of computer technology. The Pest Management program is increasing its use of computers to provide quicker delivery of information to clientele. Information which can best be provided by computers includes that on registration, recommendations, and uses of pesticidal chemicals. Computers are also used to recall information on crop development, pest outbreaks, weather data, and management techniques for most efficient use of pesticides. The program is developing innovative systems that assist clientele in decision-making for the best protection of their crops.

A second important educational vehicle is the provision of written material. Technical and popular publications provide information on the importance and management of beneficial and pest insects, pathogens, nematodes, and weeds on crops. Publications on pest species affecting animal and human health are also provided. The program uses advanced educational tools, such as IPM Manuals (Anonymous 1982, 1984) which provide state-of-the-art information on integrated pest management for a specific crop. The manuals contain a general discussion of crop biology and production practices in relation to pest damage, seasonal monitoring charts and guidelines, biologies and IPM guidelines for each key pest, and quality photographs for identification of pest and beneficial species. Increasingly, the Pest Management Program is moving to the use of audio-visuals. For example, videotaped presentations are useful for training growers, pest control advisers, and other clientele in such areas as sampling techniques, status of current research, and methods of pesticide application and safety.

In addition, demonstration plots have long been used to show clientele methods of sampling and pest identification, varietal response, and crop and pest management strategies. Most trials are usually conducted in the field in cooperation with growers. Some are better conducted on University-controlled experimental ground. Likewise, training is provided by program specialists to county-based farm advisors, via field days or classroom-style seminars.

Interaction of specialists and advisors with personnel responsible for pesticide control is another form of educational activity for the program. The goal is to interact with other county, state, and federal agencies, pertaining to the use and regulation of pesticides. The Pest Management Program maintains current information on pesticide registrations, which is in turn available to specialists, county advisors, growers, private sector pest control advisors, and the public. Computer delivery of this information is now available on a number of pesticides and commodities; the data base is currently being expanded.

Monitoring and Evaluating Pest Problems. The job of monitoring and evaluating pest problems minimizes losses while lowering costs. The program anticipates using about 20% of its resources in this area. In the past, the majority of resources allocated to this general category were in monitoring, sampling, and diagnosis. These activities have been identified, however, as being of lower priority; emphasis will be placed instead on developing economic thresholds and pest and host phenology models. As greater priority is given to the development of action (economic) thresholds, resource allocation is expected to double. Action levels can be developed for pests (by computer model) to indicate when a control program should be initiated.

Computers used in development of pest and host phenology models are a priority for Pest Management. One example is heat-driven (day-degree) models providing developmental times for both pests and host plants. When used to forecast host development and pest outbreak, they provide a useful management tool (see Sect. 2.4.2.3, Table 3, which presents developmental thresholds and day-degree sums for nine pests in California).

Methods of Management. Testing of new compounds and development of new uses for existing compounds has traditionally been a high priority. With new technology, research, and public concern over use of pesticides, this has been identified as a lower priority. Emphasis is being placed on current and potential techniques, so pest populations can be managed, economic crop damage avoided, and adverse side effects minimized. These techniques include various methods of biological control, cultural control (including host resistance), and new technology. Methods of control must be evaluated to assure that the best management tactics are being utilized. Equipment and techniques will be refined to allow placement of toxicants in the micro-environment of the pest.

Environmental and Economic Impact of Pest Management. The amount of time and resources allocated to environmental concerns and economic pest management is expected to double in the future to approximately 15% of the total. There is an increased awareness of the program's responsibility to the environment, health and safety, and socio-economic factors.

Pesticide safety is an increasingly important area, and in the future more information will be provided on the toxicology of pesticides and the epidemiology of poisoning. The Pest Management Program currently offers training programs on pesticide safety in both English and Spanish, as a large percentage of agricultural workers are Hispanic. Additional educational and resource programs are needed to inform the public on safe methods of pesticide use in agriculture, forestry, and urban public health.

Impact of pesticide application on soil, water, and air quality is also an area of concern for the program; resources allocated to this area are expected to triple in the future. The development of resistance by pests to pesticides is an increasingly important problem. Program resources allocated to solutions in this area are expected to double. A related problem, secondary pest outbreaks, also will receive additional attention. As resources become more scarce both within the University and without, the financial impact of pest management practices becomes increasingly important to consider. Work in the areas of cost analysis of pest management practices is expected to more than double.

Special Projects. Two specific projects led by Pest Management specialists are also identified as program priorities, and receive special funding. One special project is that of mosquito research. The goal of this program is improved materials, methods, and strategy for mosquito control for the protection of animal and human health. A second project is the Western Regional Pesticide Impact Assessment Program, which evaluates pesticides in regard to their risks to people and the environment.

3.4.4 Availability of Auxiliary Services

UCCE is different from its counterparts in many states in that it does not operate soil and plant clinics where farmers and urban citizens can have routine fertilizer and nematode analyses of soil carried out for a modest fee. Nor are central plant clinics operated for disease and pest diagnosis, although many problems that cannot be routinely diagnosed by an advisor are sent to the specialists for diagnosis. UCCE does operate two soil and plant tissue laboratories where farm advisors and specialists working in demonstration plots and/or applied research can request analyses of soils or plant tissues, including livestock feeds.

There are a number of private laboratories for soil and nematode analysis in California. The University encourages this development and provides diagnostic tools and training for private laboratory personnel. Specialists involved in the solution of field problems do provide some routine diagnostic services. However, UCCE has not had a history of highly organized monitoring networks for pests and diseases. Instead, the California Department of Food and Agriculture (CDFA), a state agency, working in conjunction with County Agricultural Commissioner (CAC) personnel would be the agency which carries out standardized routine monitoring services. This would include such items as ongoing trapping for fruit flies and Japanese beetles, *Popillia japonica.*

Annual surveys are carried out for beet leafhopper, *Circulifer tenellus*, the vector of curly top virus; and pink bollworm, *Pectinophora gossypiella* of cotton in the San Joaquin Valley.

Often these surveys are carried out in a cooperative effort between national and state agencies. Finally, periodic surveys are carried out by the California Department of Food and Agriculture for many pests such as the cyst nematodes, *Globodera* spp., of potato.

Two meteorological services are presently provided statewide. One is the fruit frost warning system provided by the Weather Service of the United States Department of Commerce. In addition to specific frost warning they provide additional supplementary information, e.g. on rainfall, that will assist farmers in making decisions on agricultural practices, such as spraying for certain fruit diseases. Farm advisors and specialists have access to this data for their daily broadcasts on radio/TV or for newsletters.

Within the UC Statewide IPM program there is an agricultural weather program. Daily weather parameters from more than 50 locations in California are fed into a computer. In addition, 30 years of historical weather information from a number of agricultural locations is summarized in the computer. This information can be accessed by farm advisors and specialists to drive plant or pest phenological models for predictive purposes or for ongoing research activities. Finally, private entrepreneurs have developed meteorological services that provide growers with specific information needed to make pest management and other agricultural decisions.

3.4.5 Administrative Structure

Figure 3 presents a diagram of the administrative structure of cooperative extension in California. Overall supervision of all extension work is in the hands of the Director of Cooperative Extension, who delegates the actual running of extension programs to the Associate Director-Programs. The latter coordinates the work of eight Program Directors in various spheres of extension work, and one of these directors is responsible for the pest management program, including its planning, staffing and the allocation of resources. He supervises the 28 campus-based specialists in entomology, plant pathology, nematology, weed science, pesticides and integrated pest management.

A parallel chain of command exists for county farm advisers, who answer to county directors and, through them, to one of four regional directors. The latter are administratively responsible to the Director of Cooperative Extension.

There are thus essentially two parallel structure, one that deals with campus-based (for the most part) specialists, and one that deals with county-based advisers. Needless to say, there are clear lines established for administrative and programmatic coordination between these two groups. Thus IPM advisors may be located at county sites where the County Director provides space and support for them, but they are responsible programmatically to the Program Director for Pest Management. Constant contacts in crop protection are also due to the fact that county-based advisors in agronomy, horticulture, viticulture, and vegetable crops are heavily involved in pest management related program activities.

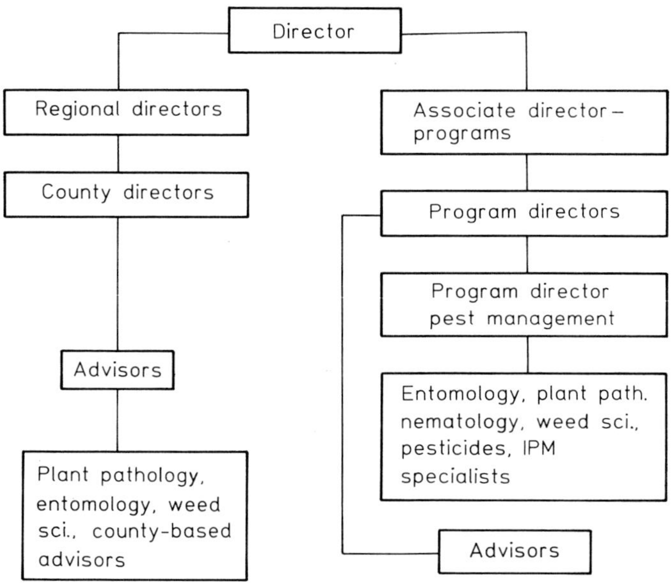

Fig. 3. Organization of Cooperative extension and of its crop protection component

In planning and implementing CE programs, ongoing communications takes place at all levels. In addition, advisors and county directors work most closely with clientele in the field, while specialists and program directors work closely with University researchers. Together they bridge the gap between University research and clientele needs. It is through this organizational matrix that the mission of Cooperative Extension can be best carried out.

3.4.6 Personnel and Resources

Personnel. It is the policy of UCCE to hire candidates with Doctoral degrees, and preferably some work experience, for campus-based specialists positions, and Master's degrees for county-based farm advisor positions. Pest management farm advisors hold college degrees in entomology, plant pathology or weed science. All specialists and advisors are academic employees of the University of California.

Advancement in UCCE is on merit. Candidates are reviewed regularly for in-grade salary adjustments. Promotions from Assistant to Associate Agriculturist, and from Associate to Full Agriculturist are major steps and require in-depth reviews. Promotions are based on both peer and administrative review of a candidate's program contributions.

UCCE Pest Management personnel are encouraged to maintain their professional standing by membership in professional societies and attendance at professional meetings. Specialists regularly attend national meetings of scientific societies and many attend international meetings. Farm advisors and specialists both attend local (California) meetings of growers' associations (e.g., the California Beet Growers' Association) and national grower organization meetings (e.g., National Cotton Growers' Council), and actively participate in the scientific portions of those meetings. Active participation as officers in these various professional organizations is encouraged and taken into consideration during review for salary action.

Resources. Cooperative Extension derives funds from four major sources; state appropriations, federal appropriations, county appropriations, and private funds, as follows:

Federal appropriation (USDA)	20.8%
Federal contracts and grants	1.0%
State of California	60.4%
County governments	13.2%
Nongovernment funds	4.6%

Total funding for the Pest Management Program, including academic and nonacademic salaries, is in excess of \$3 million. The salaries of academic staff located in county offices are paid from state and federal funds by the University of California. Their support, such as secretarial services, field assistance, travel, supplies, and other expenses, is provided by cooperating county governments.

Several major projects and needs are supported by grower groups through commodity marketing orders. Growers assess themselves a fixed sum per unit of production to support research, implementation, and promotional activities.

Several major projects are also supported using a method similar to the proceeding by the Almond Board of California, California Tree Fruit Agreement, Fresh Market Tomato Advisory Board, and Iceberg Lettuce Research Advisory Board. This support given to UCCE personnel by grower groups indicates that farmers are satisfied with the service provided by UCCE and willing to spend money in addition to general tax funds for the special programs of UCCE.

3.4.7 Summary

The University of California Cooperative Extension Service is dedicated to identifying and solving problems for the people of California. To support CE's programs, funding is obtained from federal, state, and county governments as well as non-government sources. Within this framework, the Pest Management Program has as its goal the integrated management of pests, including insects, weeds, nematodes, pathogens, and vertebrates, which impact upon California's agricultural communities and urban environments. In meeting this objective, campus-based specialists and county-based farm advisors work to bridge the gap between University research and the clientele CE serves. Priorities are established and resources allocated through a planning process involving input from each of these areas. It is this working relationship which enables Cooperative Extension and the Pest Management Program to bring about the changes which will best serve clientele and the state of California.

References

Anonymous (1982) *Integrated pest management for tomatoes.* Prepared by the IPM Manual Group of the Statewide IPM Project, University of California, Davis. M. L. Flint, Director. P. Rude, Senior Writer. University of California Agricultural Science Publication No. 3274

Anonymous (1984) *Integrated pest management for citrus.* Prepared by the IPM Manual Group of the Statewide IPM Project, Univ Calif, Davis, Univ Calif Agric Sci, Publ No 3303, 144 p

Flint ML (ed) (1981) Annu Rep Univ Calif Statewide IPM Project, 1980–81. Univ Calif IPM Project, Davis

Meyers J (1982) Cooperative extension long-range planning statement, Univ Calif Coop Ext, Berkeley, 130 p

3.5 Dissemination of Pest Management Information in the Midwest, U.S.A.

B. D. BLAIR

3.5.1 Sources of Information

Sources of pest-related research results used to answer farmer problems and questions are numerous in the midwestern states and because of this, those involved in information dissemination are numerous. Many farmers in midwestern states and most agricultural suppliers' personnel have one or more college degree. The formal education level of farmers and farm supply personnel enhances the educational process required for updating farmers and their suppliers' personnel in the Midwest. Information gets to the farmer in many ways involving people in both public and private employment. Many educational programs for farmers result from the combined efforts of plant pathologists, entomologists, weed scientists, and others with pest information, including those interested in rodents and birds. Midwestern farmers frequently obtain information on the same subject from several sources before making decisions.

Agricultural experiment stations associated with land grant universities are the source of much data and many publications needed by midwestern agricultural educators. Researchers often obtain information concerning pest problems from educators and farmers, thus allowing research results that are meaningful and helpful to educators and farmers. Research data and publications from the USDA Agricultural Research Service and non-experiment station associated personnel supplement that supplied by experiment stations.

Research findings needed for use in production management in increasing amounts and of increasing importance are being supplied by pesticide manufacturers and distributors, who also generate information needed in pest control. Private consulting firms are a relatively new development in the Midwestern states, in contrast to other parts of the United States, where consultants have been available for many years. Consultants through contract research with chemical manufacturers and distributors are providing an increasing amount of data.

Part of the annual research results are evaluated and published as pest control recommendations yearly by both those in the public and private sectors throughout the Midwest. Control recommendations usually serve as the foundation for educational programs involving agricultural supply personnel, government employees, and farmers.

3.5.2 Dissemination of Information by Public Agencies

The Cooperative Extension Service as a part of the land grant university provides much educational program leadership in pest-related fields. Extension specialists are usually housed on university campuses and develop pest control recommendations and disseminate this and other pest-related information to farmers in many ways. Each county in the Midwest has a county agricultural agent who is responsible for agricultural education programs in the county. Extension specialists provide in-service training for county agents as well as serve as resource people for educational meetings. Extension personnel use mass media including radio, television, newspapers and newsletters. The use of phone calls and office visits play a part as information dissemination methods. Recently developed pest management programs have included the use of computers and telephone answering devices. These two developments have allowed weather and other important and timely information to be included for pest management programs where needed on a daily or multi-daily basis.

In recent years, extension personnel in the Midwestern states have developed pest management programs for several field crops including corn, alfalfa, wheat, and soybean. These programs involve training and supervision of personnel to supervise and to scout farm fields. Fields are usually scouted weekly for insects, weeds, and diseases, and the weekly reports are provided to farmers who pay for the scouting. Extension Service personnel are usually responsible for pest survey programs that include insects, weeds, and diseases. Survey reports are usually disseminated weekly. Survey reports serve as the base information for pest management programs.

Many personnel from governmental agencies other than universities are involved in the dissemination of pest management-related information and in conducting educational programs necessary to get the information to the user, the farmer. Both federal and state regulatory agency personnel are involved in preparation and delivery of educational materials that are associated with pest control. These materials commonly are associated with pest control programs that usually involve area regulatory actions and are sometimes associated with quarantine programs.

Vocational agriculture teachers are involved in agricultural education with high school students and with adults as well. Many vocational agriculture teachers serve as advisors to organized young farmer groups in local school districts, and these groups sponsor agricultural tours and other educational meetings for group members.

Most counties in the Midwest have Soil Conservation Districts that are associated with the Soil Conservation Service unit in that county. Many districts have employees and are governed by boards composed of elected members. These farmer board members and employees plan and conduct educational meetings usually in cooperation with other public and private groups.

3.5.3 Advice Disseminated by Suppliers and Consultants

Personnel of agricultural manufacturers are playing an ever-increasing role in pest-related educational programs and in dissemination of pest-related information. Many excellent pest identification publications, including color photographs of weeds, insects, and diseases have been published by those in the agricultural supply business.

Many colored slide sets, movies, and leaflets that aid in teaching identification, biology and control of weeds, insects and diseases have also been developed by farm supply personnel, who also conduct educational meetings throughout the Midwest. These include short course-type meetings for farmers where a specific pest is the only meeting subject, meetings where all major pests of a particular crop are covered, and the use of the open house approach. Open houses allow agricultural suppliers to set up booths where products are displayed and allow farmers to question suppliers about products. Formal educational meetings usually are associated with the open house approach. Agricultural manufacturers frequently sponsor seminars and meetings for distributors and dealers as well as farmers. Dealers of agricultural supplies are the last "link in the chain" of education, since they are usually the last contact made by the farmer before farmer action takes place. Dealer establishments provide educational materials, phone and office consultation, newsletters, and recorded phone messages for customers.

More recently, private consultants have become a major force in the delivery of pest management information to growers in several areas of the Midwest. This trend is expected to continue. Consulting programs usually involve a total crop management program, with pest management being an important program segment.

Educational efforts by consultants include the conducting of educational meetings for their customers, newsletters to customers, scheduled farm visits, and supply of educational materials by mail, phone and computer.

Midwestern pesticide applicators are involved in the information flow and in many instances serve the same educational role as does the dealer of agricultural supplies. Applicators frequently sponsor infield meetings during problem periods.

Since many are involved in getting information from the research community to the grower, the dissemination system may appear somewhat complicated. Many educational programs are the cooperative effort of public and private employees and progress in farm production in the Midwestern section of the United States would indicate the cooperative educational efforts of many result in success.

3.6 Implementing a Statewide Pest Management Program for Texas, U.S.A.

R. E. FRISBIE and G. M. MCWHORTER

3.6.1 Introduction

Texas is geographically located in the central, southern United States. It is ecologically rich and diverse, possessing nine distinct vegetative zones. Geographically, it spans from 26° to 36° northern latitude, having a subtropical climate in the southern part of the state and a semi-arid climate in the central and northwestern portion of the state. Far west Texas is considered arid and far east Texas is classified as humid. Rainfall ranges from 200 mm per year in far west Texas to near 1500 mm in east Texas.

Agriculture is a major state industry. There are approximately 55.4 million ha used for farming and ranching in Texas. Of the 11.4 million ha of cropland under cultivation, about 3.2 million receive irrigation. Nearly 65% of all irrigated acreage is located in the High Plains of northwest Texas. In contrast to many agricultural producing states, the majority of crop acreage is grown on upland prairies, as opposed to river bottom agriculture.

The variety of crops produced over diverse geographical areas and under a wide range of ecological and economic conditions forms the backdrop for integrated pest management (IPM) programs in Texas. Although Texas has many distinct agricultural production regions, for organizational convenience and purposes of this discussion, we have divided them into the nine areas indicated in Fig. 1.

Insects, weeds, and diseases are recognized as major limiting factors in the production of crops in Texas. For example, insects cause losses of nearly $100 million annually to the state's cotton crop, while insecticide treatments increase production costs by another $20 million (Frisbie and Adkisson 1981). The guiding principles that have dictated the need for implementing IPM programs in Texas were born from a series of insect-related crises. In general, the reliance on a single tactic to control insects, i.e. insecticides, was prevalent throughout Texas and much of the cotton-producing areas of the United States throughout the 1950's and 1960's. The boll weevil (*Anthonomus grandis*) first developed resistance to the organochlorine insecticides in the 1950's, followed by the tobacco budworm (*Heliothis virescens*). The resistance of the tobacco budworm to all known insecticides occurred in the late 1960's, causing widespread alarm across much of the cotton-producing areas of the state. This was the turning point that truly initiated the change from insecticide dependence to a multi-tactic approach for economic suppression of a key insect pest species. From the problems encountered with the tobacco budworm, and to a certain extent the bollworm (*Heliothis zea*), research and extension efforts were organized based upon ecological and economic principles. Careful field monitoring was recommended, along with timed application of insecticides based upon well-defined economic thresholds. Perhaps the greatest breakthrough came in non-insecticidal approaches for the suppression of key pests. These included cultural tactics such as time of planting and destruction of post-harvest crop residues, controlled uses of water and fertilizer to make the plants less attractive to insect pests, development and use of short-season, rapidly fruiting cotton cultivars designed to escape the late-season pest populations, an awareness of naturally occurring parasites and predators, the value of pheromone traps for monitoring, and the use of a computerized pest simulation model, Heliothis MOTHZV developed by Hartstack and Witz (1983). The catastrophe that occurred in cotton in the late 1960's almost occurred in sorghum in 1974–75, as the greenbug (*Schizaphis graminum*) became resistant to most insecticides. Fortunately, greenbug-resistant sorghum hybrids were available and were planted widely

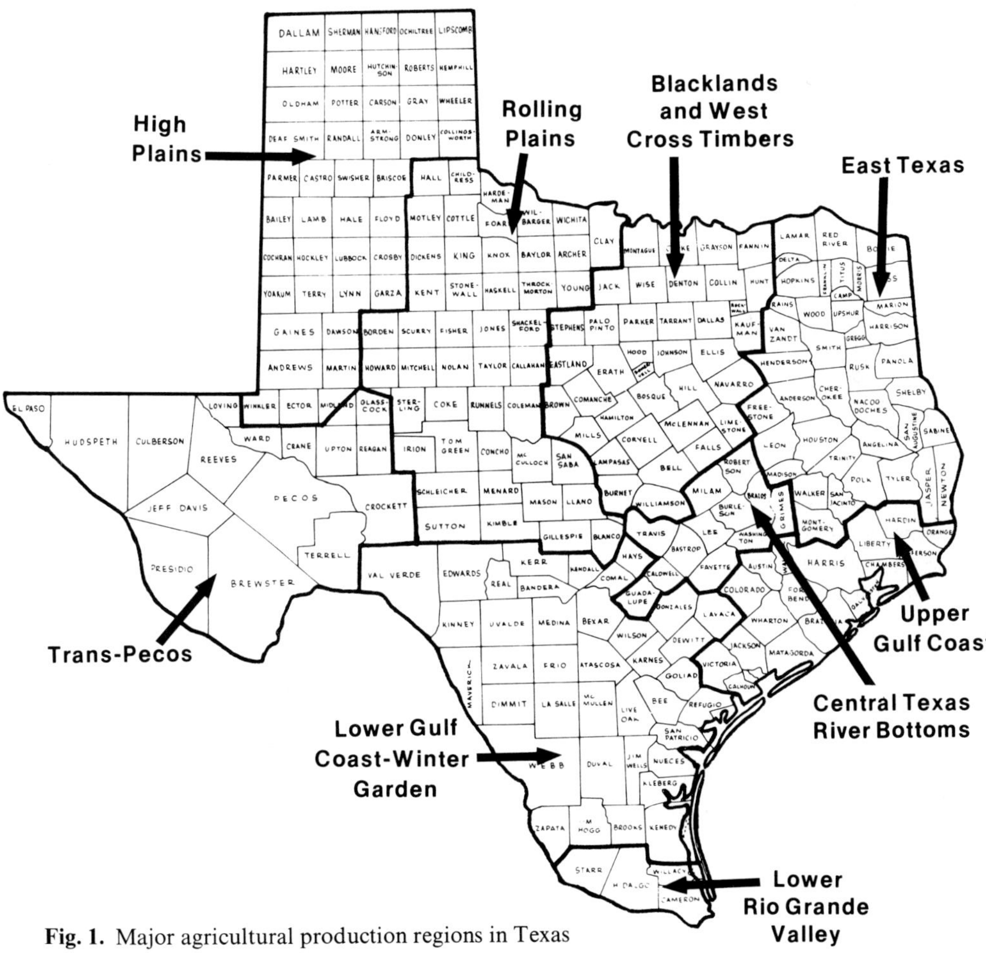

Fig. 1. Major agricultural production regions in Texas

before major damage was sustained. A system was also developed to integrate the management of sorghum insect pests with cotton pest management to minimize destruction of the beneficial arthropods that live in sorghum and migrate to cotton.

The initial emphasis on insect pests as the major focus of concentration for IPM programs began to shift in the late 1970's. The importance of managing multiple pests including weeds, plant pathogens and nematodes, again reshaped the philosophy of IPM in Texas. Management of pests must be placed in the proper perspective, i.e., as they relate to the growth of a healthy and productive crop. This system represents the future philosophical trend of pest control in Texas.

Emphasis was first placed on cotton and sorghum production systems for the implementation of IPM programs through the Texas Agricultural Extension Service. Other crops such as peanuts, corn, rice, soybean, wheat, alfalfa, and the tree crops, pecan and citrus, were included in the late 1970's and early 1980's in what was termed as "The Texas Statewide Pest Management Program". A crop-by-crop discussion of each of the previously listed IPM programs will not be presented in detail. Rather, an overview will be presented and the common features of each of these programs will be identified. There are several features, regardless of crop, that are common requisites in order to implement IPM at the farm level. These proceed in a stepwise fashion, as discussed by Adkisson et al. (1981): (1) research must be done to develop, test, and evaluate the IPM package in experimental plots; (2) field surveys must be made to determine the best locations for implementing the package and for making adjustments to local conditions, producer needs and attitudes; (3) extension specialists should conduct on-the-farm demonstrations of the IPM package, making adjustments for local management; (4) the package should be tested by the most respected and innovative farmers in the area; and (5) economic evaluations of farm demonstrations and farmer tests should demonstrate the profitability of the package. It is of utmost importance, if an IPM system is to be adopted, that it be highly practical and easily included into farming practices. If an IPM tactic requires any shifts or major changes in farming practices, it will most likely result in farmers rejecting the technology. It is the responsibility of extension specialists to present new technology in as simple a form as possible in order to gain immediate acceptance by farmers who are most likely to adopt this new technology. As a part of the educational process, it is essential to draw upon previous economic evaluations that show the potential profitability of the technology. Profitability in terms of increased net profits, either by reducing variable costs or increasing yields, is a primary consideration for adoption. Profitability, however, is not the only driving factor in farmer adoption. Risk also must be considered. If a new IPM tactic or strategy either performs, or is perceived to perform, with a higher degree of variability than the conventional practice, adoption may not take place, or will take place at a significantly slower rate.

The following sections include a description of the Texas Statewide Pest Management Program. The planning process for IPM will be reviewed. A statewide insect management plan for cotton and sorghum was first developed in

1974 (Frisbie and Adkisson 1974). This plan was later revised and considerably expanded in 1981 to include multidisciplinary IPM research and extension objectives on most major crops in the state (Frisbie 1981). The following discussion will also emphasize program organization, administration and staff training. Involvement of farmer organizations and the role they play in supporting the implementation of pest management programs in Texas will also be included. An overview of some of the unique tactics and strategies that have been adopted by farmers as a result of the Texas Statewide Pest Management Program will be reviewed.

3.6.2 A Statewide Pest Management Program for Texas

During the 1960's, agricultural producers in Texas began to experience the economic and biological repercussions of being too solely dependent on insecticides to control cotton insect pests. Concern that pesticides were contributing to the deterioration of environmental quality caused governmental agencies, agricultural producers, and the general public to "reassess" the wisdom of using insecticides in an often biologically unsound manner. As previously discussed, confronted with this complex problem, agricultural producers and major regional cotton commodity organizations requested that Texas A & M University begin to develop a new "integrated" approach to cotton insect control. Improper insecticide use was also producing chemical resistance in some pests and greatly increasing production cost and eroding net profits. The Texas Agricultural Extension Service and the Texas Agricultural Experiment Station worked together to formulate new integrated insect management systems in response to producers and their representative organizations. The interaction between producers and their land-grant university established a partnership that focused on improved pest management techniques.

To accomplish the objectives of the newly founded integrated insect pest management program, an organizational framework and a long term program implementation strategy had to be developed. Research scientists, extension educators, and agricultural producers met to identify technology needs. Producer accessibility to an extension IPM professional was identified as the primary need for the successful implementation of IPM tactics. The agricultural extension service established an organizational framework that would provide a trained extension staff member for each IPM program area. Single or contiguous multi-county areas, where producers expressed a desire to support and use the pest management program, were identified. In 1972, a pilot stage IPM program was established in each of four areas; the Lower Rio Grande, Blacklands, Winter Garden, and Far West Texas. Regional cotton and grain commodity associations served as sponsors for these new programs. These program locations provided an opportunity for IPM program tactics and educational methods to be tested over a variety of agricultural production systems and insect pest situations. An extension IPM professional (County Extension Entomologist) was hired for each program and was given the primary responsibility for introducing the IPM concept to producers. This staffing and organizational arrangement proved to be highly successful and has been identified as the primary factor for the significant adoption of the IPM philosophy in Texas agriculture. As a result of the early successes experienced in the pilot programs, agricultural producers requested additional programs be established throughout the state. In the ensuing years, additional staff have been hired and programs initiated in over 70 Texas counties. The IPM program is evolving into a multidisciplinary educational effort, assisting and training producers of cotton, grain sorghum, corn, citrus, peanut, pecan, soybean, rice and wheat.

As the extension IPM program began to materialize and expand, an administrative and technical organization was developed that would provide management and accommodate growth in the new program. Figure 2 presents the current administrative and technical organization of the extension IPM program in Texas.

Extension Agents-Pest Management (formerly County Extension Entomologists) trained in specific crop protection disciplines, work directly with agricultural producers and other members of the county extension staff to achieve IPM program educational objectives. Extension Agents-Pest Management become thoroughly familiar with the agricultural pest management needs of producers in their geographical area. These extension educators have the following tasks: (a) assist producers in monitoring pest problems and evaluating control management alternatives; (b) assist in the recruitment, training and supervision of field scouts and scout supervisors employed by participating farmers; and (c) are responsible for the planning, implementation, evaluation and interpretation of results of demonstrations established in the pest management program area. County-based Extension Agent-Pest Management staff are assisted by a group

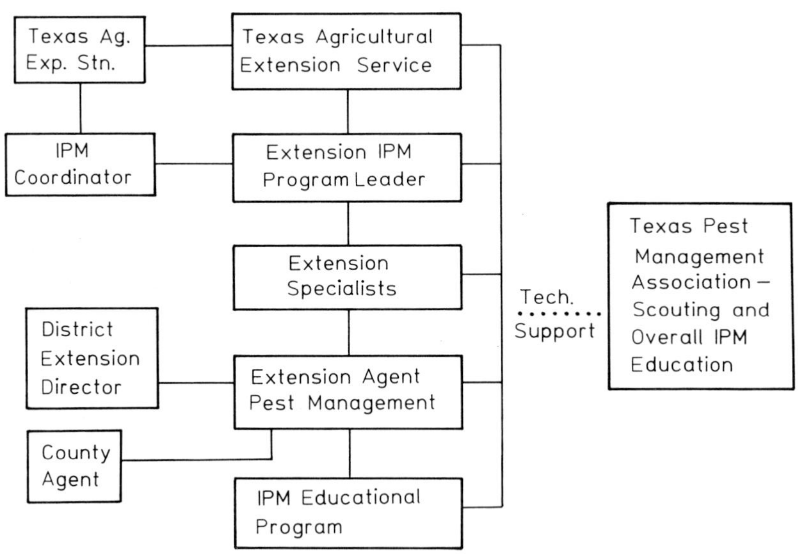

Fig. 2. Texas A & M University System organizational structure of the Texas Statewide Pest Management Program

of highly trained extension specialists headquartered at regional Texas A & M University Research and Extension Centers.

Extension specialists have usually completed the Ph. D level of education and are trained in a variety of agricultural disciplines. They serve as the primary technical support group for all county-based educational programs. Extension specialists provide technical training and are also involved in presenting educational information directly to agricultural producers. The extension specialist staff provides a comprehensive, multidisciplinary task force that assists agents in all phases of program implementation.

The *Extension IPM Program Leader* is responsible for overall technical program organization and leadership and works closely with extension administration and specialists in the planning, implementation and evaluation of the educational program.

The IPM program leader also serves as the primary budget planning and management officer for the program. Fiscal support for the Texas Agricultural Extension Service IPM program comes from federal and state sources. Funding provides IPM staff salaries, travel, and program equipment needs. By working cooperatively with extension specialists, fiscal officers, and extension administration, an annual budget is developed that supports the educational activities of the Statewide Pest Management Program. The IPM program budget and plan of work is annually reviewed and approved by the Director of the state and Administrator of the Federal Extension Service of the U.S. Department of Agriculture.

In 1979, the position of University *IPM Coordinator* was established to further enhance communications and liaisons between the academic departments and research and service agencies of the Texas A & M University System. The University IPM Coordinator also serves as the resource person for the Extension IPM Program Leader in planning and evaluating educational activities.

The extension IPM program has greatly increased its scope of educational activities designed to assist producers of major agricultural commodities throughout the state. Although the original emphasis was placed on the management of the boll weevil and bollworm in cotton, a multiple crop-pest management approach is now apparent. The crops and major pests now included in the extension IPM programs are listed in Table 1.

Considerable progress has been made in the development of multiple crop-pest management strategies. Further expansion of educational programs is planned for the future. As technology becomes available to provide interdisciplinary management of a wide variety of production-limiting pest problems, they will be incorporated into the educational programs. Increased emphasis on total crop management, including such important factors as varietal selection, cultural practices, fertility, and irrigation management, will become more evident.

Well-trained technically qualified individuals have been the key to the successful implementation of IPM in Texas. Traditionally, county-based extension IPM program staff have a Bachelor of Science or Master of Science degree in a plant protection discipline such as plant pathology, entomology, or weed sci-

Table 1. Major pests monitored in IPM programs

Crop	Pests monitored in IPM programs		
	Arthropods	Diseases	Weeds
Cotton	Boll weevil Bollworm Tobacco budworm Cotton fleahopper Pink bollworm	Seedling disease complex Root rot	Pigweed (21) Johnson-grass (22) Nutsedges (23) Silverleaf nightshade (24)
Corn	Southwestern corn borer Banks grass mite (1) Two-spotted spider mite (2)		
Sorghum	Greenbug Banks grass mite Sorghum midge		
Wheat	Greenbug Wintergrain mite (3) Brown wheat mite (4) Armyworm complex	Rust	
Pecans	Pecan nut casebearer (5) Pecan weevil (6) Hickory shuckworm (7) Pecan aphid complex	Pecan scab Powdery mildew (15) Stem end blight (16)	
Citrus	Citrus rust mite Texas citrus mite (8) Scale complex	Melanose (17) Greasy spot (18)	
Peanuts	Burrowing bug Armyworm complex Lesser cornstalk borer	Leaf spot Southern blight (19) Peanut rust (20)	
Soybeans	3-Cornered alfalfa hopper (9) Soybean looper (10) Velvetbean caterpillar (11) Rice stink bug (12) Budworm (bollworm)		
Alfalfa	3-Cornered alfalfa hopper Alfalfa looper (13) Alfalfa weevil (14) Armyworm complex		

(1) *Oligonychus pratensis,* (2) *Tetranychus urticae,* (3) *Penthaleus major,* (4) *Petrobia latens,* (5) *Acrobasis nuxvorella,* (6) *Curculio caryae,* (7) *Laspeyresia caryana,* (8) *Eutetranychus banksi,* (9) *Spissistilus festinus,* (10) *Pseudoplusia includens,* (11) *Anticarsia gemmatalis,* (12) *Oebalus pugnax,* (13) *Autographa californica,* (14) *Hypera postica,* (15) *Microsphaera alni,* (16) *Botryosphaeria ribis,* (17) *Phomopsis citri,* (18) *Mycosphaerella citri,* (19) *Sclerotium rolfsii,* (20) *Puccinia arachidis,* (21) *Amaranthus* spp., (22) *Sorghum halepense,* (23) *Cyperus* spp., (24) *Solanum elaeagnifolium.* Scientific names not listed can be found in the text

ence. Agronomists and horticulturists have also served as Extension Agents-Pest Management. In addition to having extensive biological training, extension staff are constantly encouraged to refresh and strengthen their technical capabilities. Numerous professional improvement opportunities are available throughout the year. Staff members are encouraged to participate in their respective professional scientific societies where new IPM information may be discussed. Extension IPM staff are strongly encouraged to remain current in their respective academic areas. Local, regional, and statewide workshops and seminars are also scheduled on a timely basis. These professional improvement activities enhance staff and interdepartmental communication and improve the technical capabilities of extension personnel.

Each year a variety of scout-producer training activities are scheduled. Local and regional scout schools are provided to teach pest identification and sampling techniques to prospective scouts. Producers are also commonly invited to scout training sessions so they may also be better able to identify pests and their damage and interpret scouting report forms. Timely meetings with scouts and producers throughout the season improve their level of expertise and also provide an opportunity to review pest management situations and alternative management tactics. This process, done on a repetitive basis, improves understanding and the adoption of pest management principles and practices.

3.6.3 Farmers, Organizations, and IPM Program Involvement

The input of producers and their representative organizations in planning, organizing, and implementing IPM programs is vital. Farmers have provided important leadership on local, regional, and state steering committees directing IPM programs. The Texas Association of Cotton Producers Organizations (TACPO) was the first producer group to show significant interest in the philosophy and implementation of IPM in Texas. TACPO is composed of a group of regional producer associations that are primarily interested in the welfare of the cotton and grain industry. These associations monitor legislation and marketing issues that affect the financial well-being of their member producers. TACPO was instrumental in providing organized producer support for the extension IPM program during the early years of implementation. TACPO requested the development of the first statewide pest management plan for cotton and sorghum (Frisbie and Adkisson 1974). This plan identified research and extension IPM needs on a region-by-region basis for the state. The plan served as the architectural plan for the statewide expansion of the Texas extension IPM program. As the pest management program expanded across the state, the leadership of TACPO, working closely with scientists and administrators of the Texas A & M University System, realized that another producer organization, solely dedicated to the implementation of pest management, should be formed. As a result, the Texas Pest Management Association (TPMA) was chartered in 1977. TPMA was established as a statewide, producer-operated, non-profit association dedicated toward the advancement of IPM in Texas.

One of the reasons for forming TPMA was to separate the administration of field scouting from the educational role of Extension. The new association assumed the fiscal and personnel responsibilities of the scouting program. Extension resumed its traditional role of providing technical information and educational assistance. TPMA is operated by a board of directors composed of members participating in IPM programs and supporting commodity organizations. In addition to the cotton and grain associations of TACPO, the Texas Pest Management Association also includes representatives from the Texas Peanut Producers Board, Grain Sorghum Producers Association, and the Texas Corn Growers Association. TPMA's long-term goal is to represent all commodities and respective associations that are interested in supporting and implementing IPM tactics.

After official organization and chartering activities were completed, TPMA entered into a memorandum of agreement and understanding with the Texas Agricultural Extension Service. The memorandum was signed by the Association President and the Director of the Texas Agricultural Extension Service in early 1978. The memorandum stated that each party mutually agreed to assume specified responsibilities for the advancement of IPM in Texas agriculture.

Since 1978, TPMA has become a full partner with Texas A & M University in the promotion and implementation of IPM in Texas agriculture. Headquartered at a statewide office located in Lubbock, Texas, the Association annually represents over 1000 agricultural producers who farm over 500,000 acres of cotton, grain sorghum, corn, wheat, peanuts, citrus, soybean and alfalfa. Member producers are represented by their local steering committee

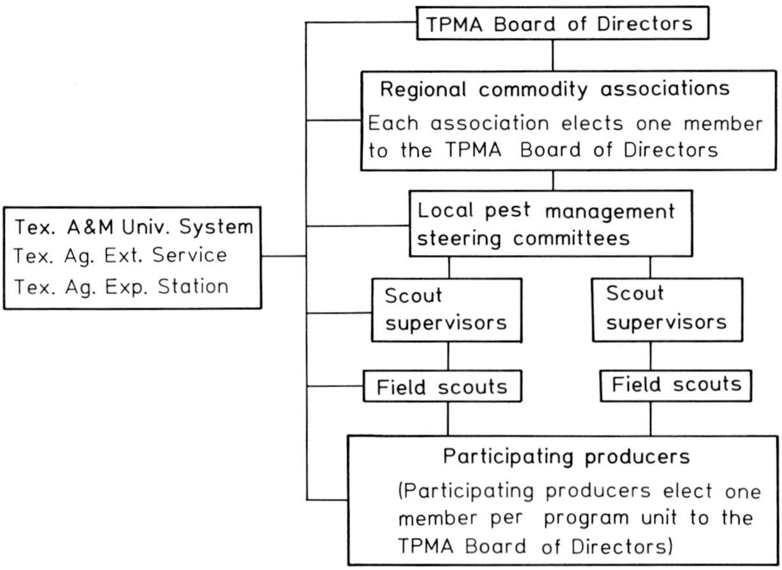

Fig. 3. Texas Pest Management Association organizational structure

chairmen and statewide board of director members. The Texas Pest Management Association has developed an organizational structure and identified goals that represent their commitment to IPM implementation (Fig. 3). The following statements describe the organization and goals of TPMA:

- Producers are direct participating members of Texas Pest Management Association and sponsored by an area commodity association. Each local pest management program consists of a county-based unit operated by a producer steering committee.
- The steering committee is comprised of producers and other interested agribusinessmen from the area and directs the operation of the scouting program.
- Producers from each IPM program are represented on the board by two directors; one, an elected participating member from each IPM program unit, and the other director is from the commodity association representing the production area. This provides statewide representation of producers on the board of directors.
- The regional commodity association sponsors and coordinates pest management efforts and assists in the development and operation of pest management units within the program area.
- There are ten regional commodity associations in TPMA representing cotton, grain sorghum, corn, and peanuts. Although not officially represented by a commodity association, citrus, wheat, pecan, rice, soybean and alfalfa are included.
- The Extension Agent-Pest Management provides technical support, serves as local coordinator for the program, and works closely with the pest management steering committee for each program.
- The County Extension Agent, as agricultural program leader for his county, plays a key role in incorporating IPM practices into the total county agricultural program.
- The Texas Department of Agriculture assists in specialized or regulatory programs for the evaluation and control of area and statewide pest problems.

The goals of TPMA are: to encourage the implementation of pest management practices; to coordinate pest management activities across the state; to acquaint producers with pest management practices; to provide a mechanism for producer-operated scouting services; to help producers organize into pest management units; to serve as a liaison between cooperating state and federal agencies; to develop state and national assistance for producers' pest management activities; and to provide a producer viewpoint on environmental concerns.

The 1974 Statewide Pest Management Plan for Texas was extensively revised in 1981 at the request of TPMA (Frisbie 1981). The revised plan included additional crops and proposed that a variety of plant protection disciplines be incorporated into a working plan for the major agricultural areas of the state. The goals of increasing farmer net profit and reducing environmental disruption remain the cornerstones of the program. The plan includes a compilation of ten subplans developed by regional teams of research and extension entomologists, plant pathologists, weed scientists, agronomists, agricultural economists, agricultural engineers and plant breeders. The plan centers around cotton, sorghum, corn, peanuts, rice and riceland mosquitoes, soybean, citrus, pecan, timber, small grain, sunflower and forage. This plan does not attempt to detail all of the pest management needs. Important areas such as livestock, vegetable and urban pest management are not discussed. Members of the Texas Agricultural Extension Service, Texas Agricultural Experiment Station, Texas Department of Agriculture and United States Department of Agriculture served on committees representing nine major agricultural production areas. In addition, the plan includes a revised review of TAMU-BUGNET with innovations and plans for the future regarding computerized information and crop- and insect-forecasting systems. Each subplan describes the agronomic and economic significance of pest management. Research needs are identified along with a listing of extension educational and personnel priorities. Techniques for program evaluation and expansion are described in the plan.

As the Texas statewide pest management program continued to increase in scope and complexity, a set of guidelines was drafted to chronicle the history of the program and describe the responsibilities of the various associations and agencies. These guidelines serve as the program's "operational manual" and are subject to periodic revision as deemed necessary by the participating parties.

The relationship established between TPMA, the Texas Agricultural Extension Service and the Texas Agricultural Experiment Station has resulted in a number of identifiable benefits to Texas agriculture. These benefits are: producers have increased net profits through IPM practices; producers play an active role in the direct operation of the local program through steering committees; field and orchard scouting personnel are provided to assist in the proper timing of control measures; assistance with scout payroll, employee records and financial management is provided to each pest management unit; assistance is provided in implementing areawide monitoring and pest suppression programs; and producers' interests are represented at the state and national level through Association activities.

The formation and development of the Texas Pest Management Association and its close working relationship with the Texas Agricultural Extension Service in the implementation of integrated pest management is a success story. The implementation of integrated pest management is best accomplished through a cooperative effort led by dedicated leaders in agricultural production and technology development.

3.6.4 Elements of the IPM Strategy in Texas

There are several elements or tactics in the production of any agricultural commodity that bear upon the suppression of pests. As a pest management strategy is developed, elements such as variety selection, resistant cultivars, and cultural practices such as crop rotation, sanitation-tillage, time of planting, fertility and water management, and the time of harvest must be taken into account. In addition, monitoring and sampling techniques should be developed or selected that most appropriately fit the pest management strategy. Economic thresholds used as guides to timing pesticide applications are a basic part of any IPM program. Crop and pest-simulation models based upon plant and pest growth characteristics and weather inputs are valuable forecasting tools in predicting future biological events. Biological control is a cornerstone of any IPM system, and its role should be examined carefully in pest suppression. Finally, regulatory programs may come into play in order to combat pests on an areawide basis. The following discussion will draw upon several crop examples to illustrate how these various elements and tactics are currently used in IPM programs in Texas.

3.6.4.1 Variety Selection and Host Plant Resistance

High quality seed of varieties that grow well under local environmental conditions is the foundation for providing a healthy plant that is able to withstand attack by a wide range of pest species.

A great deal of research and extension effort has been devoted to the development and demonstration of crops that are resistant or tolerant to insects and diseases. Host plant resistance is a major factor in preventing plant damage caused by plant pathogens. As an example, Horne et al. (1982) annually publish disease ratings for commercial grain sorghum hybrids that list resistant and susceptible cultivars. This disease rating system is used throughout the entire state, enabling farmers to select an appropriate grain sorghum hybrid for their particular region that is resistant or tolerant to one or more of the following diseases: maize dwarf mosaic virus, sorghum downy mildew (*Peronosclerospora sorghi*) or head smut (*Sphacelotheca reiliana*). The *Texas Plant Disease Handbook* provides comprehensive lists of several disease-resistant plant varieties of wheat, corn, sorghum, rice, barley, oats, peanuts, sugarcane, cotton, several vegetable crops, pecan, peaches, and grapes (Horne et al. 1983). Varieties that are resistant or moderately resistant are not only identified, but those varieties that are susceptible or highly susceptible to a particular disease are also indicated.

Insect-resistant varieties have also been identified for cotton, sorghum, wheat, rice, and other crops. Clearly, one of the major features that separate Texas cotton culture from the remainder of the cotton-growing regions of the United States is its use of rapidly fruiting, short-season cotton cultivars. For the most part, Texas produces cotton on upland prairies, as opposed to river-bottom agriculture. The "short-season cottons", as they have come to be known, have several valuable attributes in the management of insects and certain diseases. Due to the precocity of fruiting, these cottons have the ability to set and produce a crop in advance of the population buildup of insect pests such as the boll weevil, the bollworm, and tobacco budworm. With the exception of isolated acreages in the Lower Rio Grande Valley, and some in the Central, South, and East Texas, most farmers plant varieties of the short-season type. The value of short-season cottons in the management of insects, particularly the boll weevil, was first described by Walker and Niles (1971). It was further elucidated by Parker (1980) by determining that rapid fruiting offers a measure of "escape" from late-season boll weevil populations. It was shown if the plants were allowed to bloom for at least 30 days with relatively low early-season boll weevil damage, there was a high probability that the bolls of the cotton plant could escape damage from the boll weevil. In most cases bolls from 10 to 12 days of age have a good chance of escaping boll weevil damage. This work was done in south Texas and has generally held true for the central and southern part of the state.

3.6.4.2 Cultural Practices

Of all the elements that are levied against multiple pest species in nearly all crop production systems, particularly annual crops, cultural practices are used most widely and most likely have the highest value in management and suppression of pests. Among the cultural practices that have been helpful in managing pests in Texas, the following are specific examples:

a) *Crop Rotation.* The rotation of cotton with sorghum has been useful in reducing insect pests that attack cotton, such as the boll weevil. Also, losses from corn root worm (*Diabrotica* spp.) may be reduced, or in many cases eliminated, by a crop rotation scheme including soybean, cotton, and other crops that are not fed upon by root worms that attack corn (Stewart et al. 1984).
b) *Site Selection.* Site selection is a management tactic used to remove the host crop "in space" from contact with a given pest. An example of this is the planting of cotton away from overwintering habitat that is favorable to the boll weevil.
c) *Sanitation and Tillage.* This practice is used for the control of many insect, weed and disease organisms. The destruction of unharvested bolls and stalks after harvest is recommended to reduce overwintering populations of the boll weevil and the pink bollworm (*Pectinophora gossypiella*) in cotton. This is accomplished by tillage and the practice of phytosanitation to remove food supplies and overwintering refugia of these pest species.
d) *Time of Planting to Escape Pest Damage.* Depending on the location within the state, cotton can be planted uniformly early to escape late-season pests, or conversely, planted uniformly later to allow a suicidal emergence of spring infesting insect pests, such as the boll weevil. Early and uniform planting dates, using an early maturing, resistant variety, is recommended throughout the state for sorghum midge control (Morrison et al. 1983).
e) *Fertility and Water Management.* Excessive use of water and high inputs of nitrogen, in general, prolong the period where cotton and other crops may be attacked by insects and diseases. A good example, again, is when cotton is exposed to late generations of the boll weevil and bollworm insect pests. Another example, blast of rice caused by a fungus *Pyricularia oryzae*, can be

controlled, in part, by lowering nitrogen applications in combination with proper water management in fields with a history of this disease.

f) *Prompt and Proper Harvesting Practices.* Once a crop has completed growth and its fruit or seed is mature, harvest should proceed immediately. If cotton harvest is delayed by rain, boll rot caused by a complex of parasitic and saprophytic organisms that include *Colletotrichum* spp. and *Xanthomonas* spp. will cause a reduction in fiber quantity and quality.

Cultural practices that are well planned are frequently a low-cost management tactic in the suppression of pests. These practices should be factored into management decisions that are made prior to the crop production season.

3.6.4.3 Field Sampling and Economic Thresholds

Up to this point, we have focused primarily on elements such as variety selection and cultural practices that will help avoid pest incidence and minimize damage prior to and during the production season. The next step in management is appropriate identification and sampling of pest populations along with measurements taken on crop growth and development. These factors enable the manager to carefully follow pest populations and relate these populations to predetermined economic threshold levels.

Accurate *identification* of pests is essential to any monitoring or scouting program. The Texas Agricultural Extension Service relies heavily on printed material such as leaflets and brochures to help farmers identify pests. Perhaps the best example of this is found in the Disease Atlas Series developed by plant pathologists of the Texas Agricultural Extension Service by identifying diseases of sorghum (Horne and Berry 1980), peanuts (Horne 1979), corn (Horne and Berry 1978), rice (Walla 1980), and soybean (Anonymous 1979). Special publications such as *Identification, Biology and Sampling of Cotton Insects* (Bohmfalk et al. 1982) are quite helpful in providing farmers, extension specialists, private consultants, and others, with the means to identify both pest and beneficial arthropod species found in cotton. Surveying and mapping weeds are an important component of our south and east Texas pest management program. *Weeds of the Southern United States* (Anonymous 1975) along with *Common Weed Seedlings of the United States and Canada* (Anonymous 1981) have been quite helpful as a guide in proper identification of weeds that occur in several Texas crops.

Monitoring, or scouting, is an essential component of all IPM programs. The purpose is to provide timely information to farmers on the occurrence and abundance of pests, crop growth stage, damage, beneficial arthropods, and overall crop conditions. This information, in turn, assists farmers and extension specialists in making appropriate control decisions. Sampling techniques are described in virtually every insect pest management guide published by the Texas Agricultural Extension Service.

Scouting involves monitoring pest populations in time and space. The development of most scouting procedures is based on pest behavioral information that allows an individual to measure insect populations at the appropriate time and location on an individual plant and across entire fields. Regardless of the scouting technique applied, it must be tailor-made to the specific pest being monitored.

Economic thresholds have been established for many major insect pests in Texas. Since this involves observing a pest population that is truly dynamic, i.e., rises and falls periodically, there must be some method of systematic sampling on a regular basis (Bohmfalk et al. 1982). This usually requires sampling of the crop at least once a week and, in cases of intense pest pressure, twice a week or more frequently. Regardless of the sampling procedure, the individual using the technique must be certain that representative samples are taken from all parts of a field. In addition, the number of samples taken from an individual field is usually dictated by the crop and the insect pest being sampled. General rules of sampling allow for field size and variability, and changes in crop growth modified by soil and water conditions and other related factors.

Once a sample is taken, it must be recorded on an easily understandable field scouting form. Figure 4 is an example of a field insect scouting form used in cotton and Fig. 5 is the field scouting form used for multiple pest monitoring of pecan. The development of annual monitoring work schedules have been very useful in estimating the time allotted to a scout workload in a pest management program. The allocation of time required for scouting is sometimes a question whether it is done by a private consultant, a cooperative extension service scout, or the individual farmer. Questions frequently arise on the average amount of time allocated for the scouting of certain pests. Figure 6 is an example of the time allotments that could be expected for scouting pecan orchards throughout the production season.

Insect pheromone and light traps are routinely used to monitor boll weevil, bollworm, tobacco budworm, and pink bollworm in cotton to coordinate and intensify scouting during peak periods of insect activity. The numbers of boll weevils trapped in Grandlure baited pheromone traps have been used as an index to predict future damage. This trap index is used to determine the need to apply insecticides to prevent damage from overwintering boll weevils in the spring. The pheromone trap index system, as it is termed, was developed by Rummel et al. (1980). The system has been adopted as a recommendation by the Texas Agricultural Extension Service for farmer use (Drees 1984). This is perhaps the first case where a pheromone trapping index system has been used as a basis for insecticide applications. *Heliothis* spp. pheromone trap catches are also used as major inputs into a computerized bollworm-tobacco budworm forecasting model MOTHZV 3.

Economic thresholds have traditionally been somewhat static in nature and have included little information on the dynamic growth habits of the crop relative to insect populations. This is not the case today. Some thresholds now relate damage to the stage of crop development and the susceptibility of certain pest species. Probably the best example is the use of short-season cottons planted uniformly early or late, depending on the part of the state, to escape damage of the boll weevil. It is now recognized that once bolls are 10 to 12 days old they have an excellent chance of escaping boll weevil damage and the economic

PRODUCER'S COTTON INSECT REPORT

Field No. ______ Farm owner; operator ______________ Date ______ Time ______

Field Name __________ Scouts ______________ County ______

											REMARKS:
Thrips	N	L	M	H*	Armyworms	N	L	M	H		
Aphids	N	L	M	H	Bollworm larvae	N	L	M	H		
Spider mites	N	L	M	H	Bollworm eggs	N	L	M	H		
Cutworms	N	L	M	H	Beneficial insects	N	L	M	H		

*(N = none, L = light, M = medium, H = heavy)

COTTON FLEAHOPPER	SITES CHECKED	
No./100 Terminals	☐ ___ Adults 1. ___ 2. ___ 3. ___ 4. ___ ___ Nymphs 1. ___ 2. ___ 3. ___ 4. ___	
% Square Set	☐ ___ Avg. no. sqs./10 plants ___ Avg. no. fruit nodes/10 plants	
BOLL WEEVIL	SITES CHECKED	
Adults	☐ 1. ___ 2. ___ 3. ___ 4. ___	
% Punctured Green Sqs.	☐ 1. ___ 2. ___ 3. ___ 4. ___	
COTTON BOLLWORM	SITES CHECKED	Egg Color: ___ White ___ Brown
No. eggs/ ___ Terminals	☐ 1. ___ 2. ___ 3. ___ 4. ___	
No. worms/ ___ Terminals	☐ 1. ___ 2. ___ 3. ___ 4. ___	Worm Size: ___ Small ___ Medium ___ Large
% Damaged Green Sqs.	☐ 1. ___ 2. ___ 3. ___ 4. ___	
BENEFICIAL INSECTS	NONE LIGHT MEDIUM HEAVY	
Growth Stage	PRESQUARING PINHEAD MATCH-HEAD SQUARES FIRST ⅓ GR. SQ.	BLOOM/BOLL MATURE BOLLS

Fruiting Rates		Sites 1	Sites 2		Sites 1	Sites 2
	No. row ft./100 squares or bolls			No. bolls or sqs. in that distance		

REMARKS:

Fig. 4. A typical producer's cotton insect report used by field scouts in the Texas Statewide Pest Management Program

threshold, which is based on the amount (15–25%) of ⅓ grown damaged flower buds, can be considerably increased toward the end of the season as bolls in this age group accumulate. There are several other examples, such as the sorghum midge, which is only considered a pest during the flowering period of sorghum.

3.6.4.4 Pest Models and Environmental Monitoring

Computerized pest simulation models have been quite helpful in determining future pest population trends. These models assist in anticipating the occurrence of population changes and provide a means by which field monitoring can be intensified during periods of intense insect activity. Weather data, along with pest population data are necessary inputs to drive these models.

D–1197

FOLIAGE PEST REPORT

Tree*	1	2	3	4	5	6	7	8	9	10	11	12	13	14	15	16	17	18	19	20	Total Rating	Trees	Avg Rating	Action Level	Check for Warning	Check to Treat
Yellow Aphids																										
Black Aphids																										
Mites																										
Beneficials																										
Leaf Scab																										
Nut Scab																										
Other Disease (________)																										
Fall Webworm																										
Walnut Caterpillar																										
LT + D (________)																										
LT + D (________)																										
Scale or Phylloxera ( )																										

ACTION LEVELS:

Yellow Aphids	50%+
Black Aphids	2.0
Mites	2.0
Fall Webworm	1 colony/tree
Walnut Caterpillar	10 colonies/acre
Sawfly	.5
Cigar Leaf Casebearer	1.0
Pecan Leaf Casebearer	1.0
Catacola	.2
May Beetles	.5
Leafminers	1.0
Spittle Bugs	.1

NOTES:

Grower ________________

County ________________

Orchard ________________

Block ________________

Variety ________________

Scout ________________

Date ________________

Fig. 5. Pecan pest management field scouting form used to monitor insects, beneficial arthropods and diseases on foliage

ANNUAL MONITORING TIME ALLOTMENT

WM/PS – Weather Monitoring/Pecan Scab
LT + D – Leaf Tatterers and Defoliators
PNC – Pecan Nut Casebearer
FW – Fall Webworm
WC – Walnut Caterpillar
YA – Yellow Aphid
BA – Black Aphid
M – Mite
SB – Stink Bug
HS – Hickory Shuckworm
PW – Pecan Weevil
PHY – Phylloxera
Sc – Scale

	March		April		May		June		July		August		September		October		November		December	
WM/PS		.5 .5	.5 .5	.5 .5	.5 .5	.5 .5	.5 .5	.5 .5	.5 .5	.5 .5	.5 .5	.5 .5	.5 .5							
LT + D		1.6 1.6	1.6 1.6	1.6																
PNC					1.9 1.9	1.9 1.9		1.9 1.9	1.9		1.9 1.9	1.9								
FW + WC						1.9		1.9	1.9		1.9	1.9		1.9	1.9					
YA							1.6		1.6	1.6		1.6	1.6							
BA									1.6		1.6	1.6		1.6	1.6					
M									.6		.6	.6		.6	.6					
SB										1.25 1.25										
HS											.25	.5 .25							1.5	
PW											1.0	1.0 1.0	1.0 1.0	1.0 1.0	1.0					
PHY						1.6														
Sc																				1.9
TOTALS		2.1 2.1	2.1 2.1	2.1 .5	.5 4.3	4.3 4.0	.5 2.1	2.4 4.3	4.0 4.6	1.75 3.35	4.6 5.55	5.5 5.85	1.5 3.1	5.1 1.0	1.0 4.1				1.5	1.9
Per Two Weeks		4.2	4.2	2.6	4.8	8.3	2.6	6.7	8.6	5.1	10.15	11.35	4.6	6.1	5.1				1.5	1.9

Annual Requirement = 87.86 Hours At $5.00 Hour = $439.00/50 acres On 50 Acres = $8.78 /Acre/Year

Fig. 6. Typical time requirements for scouting insects and mites of pecan (Cooper et al. 1983)

The major thrust of computer use in the Texas Statewide Pest Management program has been under the overall program title of BUGNET (Jackman 1981). This program consists mainly of insect prediction models, the most frequently used being MOTHZV, a bollworm and tobacco budworm model, a pecan nut casebearer (*Acrobasis nuxvorella*) model and a southwestern corn borer model. MOTHZV is a combined insect/plant phenology model. It is heat-driven and models the insect development as the crop develops. Pecan nut casebearer and southwestern corn borer models are also heat-driven, but they predict percent emergence that is empirically based on field data.

Statistical analysis of pest management program data is also involved in the computing activities. The IBM statistical package is used locally on microprocessors for data analysis. The SAS (Statistical Analysis System) language is implemented on the mainframe computer (AMDAHL) and statistical analysis of data is also performed centrally using this computer in College Station.

Other uses of the computers in the Statewide Pest Management Program are in the area of developing new predictive models. Boll weevil trap catch data is now being analyzed to determine a base developmental temperature and to develop predictive models. The predictive model for pink bollworm phenology is now near completion. It is being developed using the HEAT UNIT MODEL, which can be adapted for use with other crops and pests beside those for which it has already been programmed.

In recent years, the focus of BUGNET has changed to reflect progress in computer technology. The focal point is now the microprocessor (personal computer). We now have the capability to provide individuals with a series of crop growth-pest forecasting models for on-farm use. Extension entomology can play a major role in teaching the value and use of crop and pest models to individuals, as well as providing areawide information with generalized forecasts.

Temperature, relative humidity, and leaf wetness are combined to predict the infection periods of early leaf spot (*Cercospora arachidicola*) and late leaf spot (*Cercosporidium personatum*) (P.F. Lummus, pers. comm.). The rate of infection of these two peanut pathogens is a function of high temperature and prolonged periods of high humidity. The use of daily meteorological observations has been correlated with slow, moderate, and rapid infection increase periods. Timing of fungicide applications have been made more precise as a result of using this technology. Similar activities are under way to develop correlations between temperature, humidity, and the infection periods of pecan scab caused by *Cladosporium effusum*. This activity is in the research phase in the pecan pest management program (Cooper et al. 1983).

In addition to pest models, several cotton growth models are available. The models are based upon plant phenology calculations that require temperature and moisture input data on a routine basis. The cotton plant model will shortly be implemented at the field level and calibrated for the major cotton production areas across the state.

3.6.4.5 Biological Control

The Texas Statewide Pest Management Program emphasizes educating farmers on the benefits of naturally occurring biological control organisms such as parasites and predators. Farmers are aware of the value of biological control in suppressing insect pest populations and are taught to be very careful in the application of insecticide so as not to disrupt these natural control organisms. Most of the educational work has centered on identification and evaluation of predators and parasites in cotton and sorghum. The amount of parasitization of greenbug by the parasite *Lysiphlebus testaceipes* is monitored very closely and this information is provided directly to the grower as part of the weekly scouting information. Similarly, in cotton, the key predators have been identified and their abundance relative to bollworm or tobacco budworm is used in economic threshold and insecticide application decisions (Boring et al. 1984). Approximately 60% of the cotton in Texas is grown without requiring chemical insecticides. Therefore, native parasites and predators and physical control mechanisms such as heat, cold, and moisture are main factors in reducing insect populations. Key predators in the Texas cotton agroecosystem include the minute pirate bug (*Orius* spp.), bigeyed bugs (*Geocoris* spp.), ants (*Solenopsis* spp.), lady beetles (*Hippodamia* spp.) green lacewings (*Chrysopa* spp.), damsel bugs (*Nabis* spp.), assassin bugs (*Zelus* and *Sinea* spp.), and several spiders. The number of key predators in relation to *Heliothis* egg numbers can be used to determine an inaction level in cotton. This means it will not be necessary to treat if a certain number of key predators are present in relation to numbers of *Heliothis* eggs. Any combination of key predators, equal to or greater than the *Heliothis* spp. egg number results in a reasonably strong assurance that damaging populations of larvae will not result (Sterling 1982). If two or more key predators are found for each small *Heliothis* larva, control measures may not be needed or a microbial insecticide containing either *Bacillus thuringiensis* or nuclear polyhedrosis virus may be used (Boring et al. 1984).

The best example of biological control in the Texas program is seen in the citrus pest management program in the Lower Rio Grande Valley (Dean et al. 1983). Key pests of citrus include the citrus rust mite (*Phyllocoptruta oleivora*), chaff scale (*Parlatoria pergandii*), California red scale (*Aonidiella aurantii*), and Texas citrus mite (*Eutetranychus banksi*). The chaff scale is under near complete biological control by the parasites *Aphytis hispanicus* and *Prospaltella fasciata*. The California red scale is kept under biological control by *A. lingnanensis*. Of the four key pests of Texas citrus, red mite and citrus rust mite undergo the most intensive monitoring and management in order not to disturb the balance of biological control within the orchard. A series of potential pests that include citrus mealybug (*Planococcus citri*), the citrus whitefly (*Dialeurodes citri*), brown soft scale (*Coccus hesperidum*), cottony cushion scale (*Icerya purchasi*), Florida red scale (*Chrysomphalus aonidum*), purple scale (*Lepidosaphes beckii*), and Glover scale (*L. gloverii*) are all under biological control by several parasites (Dean et al. 1983).

3.6.4.6 Regulatory

An extensive number of plant materials that may be infested with pests are restricted from entry into the state. An example of a regulatory program affecting a major pest species is the Pink Bollworm Control Program on cotton.

3.6.5 Methods of IPM Program Information Delivery

Many methods are used in the Statewide Pest Management Program to transfer information to agricultural producers. Vast amounts of information on the status of pests and host plants are gathered by the extension coordinated scouting program. Scouts use self-carbonizing field report forms to tabulate field observations on pests and crops. Copies of the report are provided to participating producers and to county extension pest management personnel. The scouting report form is an effective teaching tool for producers to learn about the relationship of pest densities and their impact on crop yield and quality. One of the primary responsibilities of the Extension Agent-Pest Management agent is to help producers learn to interpret and use the information provided by scouting report forms.

Extension pest management staff commonly schedule "in-field" meetings with producers. These meetings allow producers to ask questions concerning pest problems and provide an ideal setting for the extension professional to teach pest identification and management principles. The Texas Agricultural Extension Service also plans and provides a variety of field days and tours that are used to review and evaluate the demonstration of IPM technology. These demonstrations are designed specifically to take proven research techniques and apply them on a large field basis so that producers might experience the technique and evaluate its usefulness on their own farms. The resulting demonstration is the primary means of adapting proven research techniques to localized situations.

In addition to tours, field days, and in-field meetings, the Texas Agricultural Extension Service has developed a wide assortment of extension and research educational publications that provide information on integrated pest management. Fact sheets, technical bulletins, pest control guides and research progress reports are updated and published by extension specialists and agricultural research scientists. These publications are printed at Texas A & M University and are distributed by extension educators throughout the state. The publications are made available to agricultural producers and any other interested individuals. Considerable manpower and economic resources are devoted to developing and maintaining a series of pest management-related publications. Many of these publications have evolved from a prior emphasis on agricultural chemicals to more emphasis on comprehensive pest management strategies. Extensive use is made of color photographs and other techniques that improve the diagnostic capability of producers and extension program personnel. Although some publications are for sale, they are normally provided free of charge to the general public. Extensive use of radio and television to transfer IPM information is also common throughout the state.

3.6.6 Evaluating IPM Programs

Texas has taken national leadership in developing methodologies for the economic evaluation of IPM programs. In addition to economic evaluations, questionnaires are routinely distributed to program participants in order to query them on their perception of the program. The purpose of this section will be to review some of the methods used to economically evaluate a pest management program and also to review some of the responses and changes that have been made as a result of grower or clientele input.

In every IPM program that has undergone an economic evaluation in Texas, increases in net profits to individual producers and aggregate increases in revenues to the region and the state have been shown. Lacewell and Taylor (1980) summarized several techniques that have been used in economically evaluating IPM programs in Texas. These include budget analysis, linear programming, and statistical analysis. Budget analysis combined with statistical analysis was first used by Frisbie et al. (1976) and Lacewell et al. (1974), when evaluations were made on Texas cotton IPM programs. Perhaps more than any other program, the cotton IPM program has undergone intensive evaluation in various production regions across the state. The Texas High Plains Boll Weevil Suppression Program was evaluated by Lacewell et al. (1974). This program was designed to keep the boll weevil from invading the High Plains of Texas, the largest cotton acreage in the state, by a carefully organized spray program in the fall to prevent boll weevils from entering diapause. The spray zone was in an eight-county area just east of the High Plains. This program, which was initiated in 1964, showed an annual, regional impact (increase in economic activity) of $63 million and a state economic impact of $82 million, where farmers benefited by a $27 million increase in net profit. An evaluation of a short-season IPM cotton production system in south Texas showed an increase in direct farmer profits of $11 million per year and an increase in economic activity for the Coastal Bend region of $94 million per year, which translated to an increase in state economic activity of $104 million per year (Masud et al. 1980). This study evaluated the results of farmers adopting short-season varieties that escaped late-season insect pests combined with intensive field scounting, the use of economic thresholds and strict cultural practices (precise planting dates and post-harvest phytosanitation). Shaunak et al. (1982) developed an economic evaluation that considered profitability and risk for the implementation of a short-season cotton IPM strategy in the Lower Rio Grande Valley. This particular study compared two sets of 3-year periods, 1973–1975 and 1976–1978. The data base was derived from the Lower Rio Grande Valley Cotton Pest Management Program conducted by the Texas Agricultural Extension Service. By studying irrigated and dryland production in short-season and conventional production systems, it was determined that the dryland, short-season IPM production system had the highest average net returns per acre, \$93.42 in 1973–1975 and \$117.31 in 1976–1978, when compared to conventional dryland systems that returned \$35.73 in 1973–1975 and \$68.06 for 1976–1978. Reduced insecticide costs and increased yield were two primary factors that affected net returns. It was concluded that by shifting to the short-season, dryland IPM production system there would be a reduction in insecticide use, lint yield would be increased, and perhaps as importantly, risk would be substantially reduced.

Frisbie et al. (1976) evaluated the pilot, extension cotton pest management program in 1973 and 1974 for three locations, the Rio Grande Valley, the Central Texas Blacklands, and the Trans-Pecos. The success of these pilot programs served as the basis for the later program expansion. The programs included the use of trained field scouts to provide growers with information on pest damage, crop stage, and beneficial insect activity in order to allow them to make more intelligent pest management decisions using economic thresholds. This educational program, as previously discussed, was highly successful because of the use of extension specialists at the county level (County Extension Entomologists), who worked with the farmers, not only during the production season, but throughout the year. The study showed that in no year did the Texas Pest Management Program fail to increase the net returns of program participants when compared to a control group of non-participants. Increased profits came from two sources: increased returns from yield and the reduction in variable costs through reduced expenditures for insecticides.

The economic implications of delayed uniform planting to control the boll weevil in the Texas Rolling Plains was studied by Masud et al. (1984). This cultural technique is used on about 800,000 acres of cotton. Over this area, cotton farmers voluntarily organize themselves to delay the planting by 10–14 days. This tactic delays fruit formation of cotton to such a point that boll weevils emerging from diapause quarters in the spring do not have sufficient feeding or oviposition sites to increase their populations. Net per acre returns are estimated at \$21.36 for the farmers using this cultural control technique when compared with farmers using conventional practices. The average annual economic impact of uniform delayed planting in the region was estimated at \$192 million. The annual economic impact to the state for this practice was estimated at \$305.19 million. Risk was also calculated. Risk estimated by coefficient of variation showed the uniform delayed planting date practice for production years of 1972–1980 was less variable than conventional cotton production practices.

A multidisciplinary (insects, weeds, and diseases) peanut pest management program was evaluated in Comanche County, Texas in 1974 (C. E. Hoelscher, pers. comm.). Increased net profits per acre of \$15.90 and \$24.42 for irrigated and dryland peanuts, respectively, were observed when compared to growers producing peanuts under conventional systems. In 1975, increased net profits reached \$ 64.00 and \$ 38.00 for these two respective cultures. Another economic evaluation was developed for the Frio/Atascosa peanut pest management program, where IPM program-participating farmer net returns were again compared with those of non-participants. Increases in yield and reduction in variable costs from pesticides were significant factors in showing that IPM program participants on the average increased net returns by \$45.93 per acre (P. Lummus, pers. comm.). Pesticide savings have increased net returns in a Fort Bend County pecan IPM program (J.N. Cooper, pers. comm.). Insecticide cost savings alone increased net profit \$1625 for a 50-acre pecan orchard.

Greenbug-resistant sorghum hybrids are grown on about 50% of the acreage in Texas. These hybrids were developed, for the most part, from Texas Agricultural Experiment Station germplasm. Conservative estimates place the value of greenbug-resistant sorghums at \$10 million per year in insecticide cost savings alone (G.L. Teetes, pers. comm.). This does not include savings from reduced yield losses. The value of sorghum midge-resistant sorghums may be \$10 million per year during normal years and \$50 million per year during severe midge damage years (G.L. Teetes, pers. comm.).

The reduction in the use of chemical pesticides has been a major reason for reducing variable costs and increasing profit. This also has been a leading factor in minimizing environmental pollution. Insecticide use in Texas has reduced dramatically since 1964. Much of this is attributable to the implementation of pest management programs (Thomas 1979). Insecticide use in cotton was at its highest, nearly 20 million pounds of active ingredient, in 1964, has dwindled to 2.5 million pounds in 1976 and has remained at that level, or slightly lower, since that time (Thomas 1979). During this same period, the total pounds of insecticides applied to cotton (and the percentage of total use on U.S. cotton) remained relatively constant (1964–76) in the southeast, increased sharply in the Mississippi Delta area, and decreased sharply in the southern plains, i.e., Texas and Oklahoma. Thus, insecticide use during these years was reduced 88.2% in Texas with a near constant acreage, decreased 13.4% in the southeastern United States with decreasing acreage, and increased 38.8% in the Mississippi Delta with only a slight increase in acreage (Thomas 1979).

In addition to the positive environmental impact of IPM programs in Texas that have been implemented over the last 15 years, certain programs such as the High Plains Boll Weevil Suppression Program that was previously discussed have prevented the need for the use of large volumes of insecticides (Lacewell et al. 1974).

Texas farmers are genuinely interested in reducing production costs by any means. IPM programs have most dramatically impacted on reductions in expenditures for insecticide. Farmers are also concerned about environmental disruption and exposure to non-target organisms. IPM programs meet both of these needs for the farmers of the state.

Texas Agricultural Extension Service personnel involved in the Texas Statewide Pest Management Program are earnestly interested in receiving input from their educational audiences. Normally, each pest management program develops a questionnaire that is distributed to farmers participating in the program. The questionnaires have proven to be invaluable in constructively criticizing the programs over the years and have allowed extension specialists to make necessary program changes for improvement. Examples of questions that are frequently included in what we term the "producer evaluation" of the program are:

1. As a producer have you found the pest management program to be useful in your farming operations?
2. Were the scouts in your opinion reliable by providing you what you considered accurate information?
3. Were the scouting reports generally useful to you in making pest control decisions?
4. Did the pest management program make you more aware of timing pest control measures in relation to insect and disease pest numbers and damage?
5. Did the pest management program make you more aware of the value of beneficial arthropods?
6. Did the pest management program provide you with more time to devote to other farming operations?
7. Was your expenditure for the hiring of field scouts to provide you with information worthwhile?
8. Do you believe the pest management program generally increased net profits, either by reducing pesticide costs or increasing yields?
9. Was the Extension Agent-Pest Management readily available to help you with pest control decisions?
10. Do you think it is reasonable for you to pay for all scouting costs if the Extension Agent-Pest Management is supported by state or federal funds?
11. Did you feel the scouts were dependable, prompt, and delivered field reports to you on a regular basis?
12. Do you feel that the pest management program was of sufficient value to you that it should be continued next year?
13. Please provide us with additional commentary that may help us improve the pest management program.

Questions such as these are either asked in direct question form with a "yes" or "no" reply or are developed as a statement asking the farmers to indicate a category such as "strongly agree", "agree", "undecided", "disagree", or "strongly disagree". Most farmers participating in pest management programs are very much interested in responding to these questionnaires. Problems have been identified such as the need for more frequent scouting, more supervision of scouts, more interplay between the Extension Agents-Pest Management and the producer, changes in number of fields and distance between fields scouted, and in some cases clarification of scout reporting procedures. Without a doubt, these questionnaires have been very valuable to the Texas Statewide Pest Management program personnel in evaluating and improving their programs in order to tailor-make them to meet farmer needs.

3.6.7 The Role of the Private Consultant

The goal of extension IPM programs in Texas is to educate growers on the advantages of strategies and tactics employed in the use of IPM. In order to meet this goal, growers are encouraged, where possible, to hire the services of a qualified private consultant. The private consulting industry in Texas has seen a rapid expansion over the last 10 years. This expansion has been due in part to the extension IPM programs that have shown farmers the need for professional assistance in making pest control decisions. Prior to the initiation of formal IPM programs in Texas in 1972, there were 15 private consultants operating in the state. Most of these consultants were entomologists engaged in field scouting for a fee. Today, there are approximately 115 private consultants practicing in the state. Consultants have expanded their services to include more overall crop management: disease and weed control, fertility, and water management. This reasonably rapid expansion has added another dimension to IPM in the state.

The Statewide Pest Management Program has provided an atmosphere in which the private consulting industry has flourished in certain areas of the state. Private consultants have formed into professional associations to serve as a common forum for their activities. There are two consultant associations in the state, the Texas Association of Agricultural Consultants and the High Plains Association of Crop Consultants.

The role of extension entomologists and other extension pest management specialists in relation to private consultants has been one of dynamic exchange. Extension IPM specialists have provided a climate for exchange of research and demonstration information. Private consultants are involved in local extension pest management steering committees and also serve on the statewide extension IPM advisory committee. Extension is proud of the fact that they have assisted several consultants in locating their businesses in areas of need. In fact, the extension pest management programs have frequently relocated to allow a private consultant to begin business.

A responsibility of the Statewide Pest Management Program is to maintain active programs within an agricultural region to serve as an index of quality for growers and consultants. It is also felt that the local extension pest management programs serve as a nucleus for the infusion of new technology.

Consultants and their scouts are invited to scout training schools. These schools are open not only to consultants, but to individuals involved in the chemical industry, and also farmers. Private consultants play an active role in Texas A & M University-sponsored pest management workshops and conferences. Also, opinions from private consultants are considered in the development of research priorities.

In the future, it is fully anticipated that there will be more private consultant businesses established across the state. The Texas Agricultural Extension Service will play a role in assisting consultants whenever possible through providing them with technical information or assisting them with contacts to begin business. The private consultant is considered part of the overall IPM effort in the state.

3.6.8 Bridging Research and Extension Programs Within and Between Disciplines

Another objective of IPM is the integration of several disciplines by combining and coordinating their activities in order to provide the Texas farmer with comprehensive technology for integrated crop management systems. The Texas A & M University System is fortunate in that it has 13 Research and Extension Centers (REC) that are strategically placed in major agricultural areas across the state. Each REC usually has a complement of plant production and plant protection extension specialists. These include agronomists, agricultural economists, entomologists, plant pathologists, and in some cases, weed scientists. It is the responsibility of these REC teams to develop an overall IPM strategy for crops grown within their region. Further, the extension specialists interact very closely with their research counterparts at the REC. They are involved in research reviews and the setting of research priorities in order that the research conducted addresses practical production problems. This system has worked exceptionally well over the years and has been a model for the development of interdisciplinary IPM programs.

Headquarters extension staff on the Texas A & M University campus are also heavily involved in interdisciplinary work. Joint leadership responsibility is shared between disciplines for the coordination of several extension IPM programs. Representation from entomology, plant pathology, weed science, agronomy, engineering, and agricultural economics is coordinated in an effort to insure appropriate interdisciplinary direction in managing the statewide program. Special research projects, such as those ongoing in cotton, sorghum, pecan, peanut, and citrus have been particularly active in joining research and extension counterparts together to develop a comprehensive program. It is the responsibility of the Extension Entomologist-Pest Management Leader to insure that extension specialists from all disciplines work together to develop a comprehensive program.

It is the role of the Integrated Pest Management Coordinator to involve research and extension interdisciplinary teams in planning both research and extension IPM programs. Coordination usually takes place on a crop-by-crop basis, with interdisciplinary teams addressing specific crop production and plant protection practices. In addition, each year an integrated pest management conference is held. It is at this time that research and extension specialists gather to share their work. This conference is open to the public and heavily attended by farmers who are actively involved in, or are anxious to learn about, new research and extension developments in IPM. The Texas IPM Conference is used as a springboard for the planning of interdisciplinary activities for the remainder of the year and into the next crop production season.

There is no easy solution for developing interdisciplinary thrusts. The success that has been enjoyed by the Texas IPM program is based on frequent communication through workshops, conferences, and special sessions among the various disciplines involved in this program. The program has been highly successful and is a credit to the great number of individuals that have willingly participated.

References

Adkisson PL, Frisbie RE, Thomas JG, McWhorter GM (1981) Organization and implementation of an integrated pest management system. Southwest Entomol 6(4):279–287

Anonymous (1975) Weeds of the southern United States. Cooperative Extension Service, Univ of Georgia Bull 761, p 45

Anonymous (1979) Soybean diseases atlas. Tex Agric Ext Serv B-1225:20

Anonymous (1981) Common weed seedlings of the United States and Canada. Coop Ext Serv, Univ Georgia, p 31

Bohmfalk GT, Frisbie RE, Sterling WL, Metzer RB, Knutson AE (1982) Identification, biology and sampling of cotton insects. Tex Agric Ext Serv B-933:43

Boring EP, Fuchs TW, Leser JF, Allen CT (1984) Management of cotton insects in the High Plains, Rolling Plains and Trans-Pecos areas of Texas. Tex Agric Ext Serv B-1209:27

Cooper JN, Johnson JD, McEachern GR, McWhorter GM (1983) Texas pecan integrated pest management manual. Tex Agric Ext Serv D-1216:3.1

Dean HA, French JV, Meyerdirk D (1983) Development of integrated pest management in Texas citrus. Tex Agric Exp Stn B-1434:15

Drees BM (1984) Management of cotton insects in south and east Texas counties. Tex Agric Ext Serv B-1204:15

Frisbie RE (ed) (1981) A statewide pest management plan for Texas. Tex Agric Ext Serv B-1378:82

Frisbie RE, Adkisson PL (1974) Summary: statewide pest management plan. Tex A & M Univ, Dep Entomol Tech Rep, p 63

Frisbie RE, Adkisson PL (1981) Introduction. In: Frisbie RE (ed) A statewide pest management plan for Texas. Tex Agric Ext Serv B-1378:10–14

Frisbie RE, Sprott JM, Lacewell RD, Parker RD, Buxkemper WE, Bagley WE, Norman JW (1976) A practical method for economically evaluating an operational pest management program in Texas. J Econ Entomol 69(2):211–214

Hartstack AW, Witz JA (1983) Models for cotton insect pest management. In: Ridgway RL, Lloyd EP, Cross WH (eds) Cotton insect management with special reference to the boll weevil. US Dep Agric, Agric Handb No 589; pp 359–384

Horne CW (1979) Peanut diseases atlas. Tex Agric Ext Serv B-1201:16

Horne CW, Berry RW (1978) Corn diseases atlas. Tex Agric Ext Serv MP-1356:16

Horne CW, Berry RW (1980) Sorghum diseases atlas. Tex Agric Ext Serv B-1323:16

Horne CW, Toler RW, Barnes LW, Fredericksen RA, Trampoda JD (1982) Disease ratings for commercial grain sorghum. Tex Agric Ext Serv B-1278:15

Horne CW, Berry RW, Amador JM, Johnson JD, McCoy NL, Walla WJ, Philley GL, Lee TA, Kaufman HW, Jones RK, Barnes LW (1983) Texas plant disease handbook. Tex Agric Ext Serv B-1140:452

Jackman JA (1981) BUGNET – A computerized pest management information delivery system in Texas. In: Frisbie RE (ed) A statewide pest management plan for Texas

Lacewell RD, Taylor CR (1980) Benefit-cost analysis of integrated pest management programs. In: Gooding EGB (ed) Pest and pesticide management in the Carribean. CICP/USAID Semin Workshop, vol II. Berkeley, California

Lacewell RD, Casey JE, Frisbie RE (1974) An evaluation of integrated cotton pest management programs in Texas: 1964–1974. Tex A & M Univ, Dep Agric Econ, DTR No 77-4

Masud SM, Lacewell RD, Taylor CR, Benedict JH, Lippke LA (1980) Economic analysis of integrated pest management strategies for cotton production in the Coastal Bend of Texas. Tex Agric Exp Stn MP-1467:34

Masud SM, Lacewell RD, Boring EP, Fuchs TW (1984) Economic implications of a delayed uniform planting date for cotton production in the Texas Rolling Plains. Tex Agric Exp Stn B-1481

Morrison WP, Hoelscher CE, Teetes GL (1983) Insect mite pests of Texas sorghum – management approaches. Tex Agric Ext Serv B-1220:25

Parker RD, Walker JK, Niles GA, Mulkey JR (1980) The short-season effect and escape from the boll weevil. Tex Agric Exp Stn B-1315:45

Rummel DR, White JR, Carroll SC, Pruitt GR (1980) Pheromone trap index system for predicting need for overwintered boll weevil control. J Econ Entomol 73:806–810

Shaunak RK, Lacewell RD, Norman JW (1982) Economic implications of alternative cotton production strategies in the Lower Rio Grande Valley of Texas, 1973–1978. Tex Agric Exp Stn B-1420:25

Sterling WL (1982) Predaceous insects and spiders. In: Bohmfalk et al. (eds) Identification, biology and sampling of cotton insects. Tex Agric Ext Serv B-933:25–31

Stewart JW, Morrison WP, Patrick CD, Cronholm GB (1984) Managing insect and mite pests of Texas corn. Tex Agric Ext Serv B-1366:19

Thomas JG (1979) Present and future strategies of cotton and sorghum pests in Texas. Pest management strategies, vol II. Work Pap Off Technol Assess, Congr US, p 75

Walker JK, Niles GA (1971) Population dynamics of the boll weevil and modified cotton types. Tex Agric Exp Stn B-1109:14

Walla WJ (1980) Rice diseases atlas. Tex Agric Ext Serv B-1181:14

3.7 Development of Crop Protection Advisory Work in Thailand within the Framework of the "Training and Visit" Approach to Extension

R. AUSHER and Y. ELKANA [1]

The "Training and Visit" (T & V) system of agricultural extension, first developed in Turkey in the late 1960's, has been applied in various forms by some 40 developing countries in Asia, Africa, Europe and Central and South America. Many countries have adopted this system in conjunction with agricultural and rural development projects sponsored by the World Bank. The experience gained in the first years of T & V extension work has recently been described in full in two publications by Benor and Baxter (1984) and Benor et al. (1984).

In 1977, the World Bank agreed to aid the Thai government in setting up an agricultural extension network, using the T & V approach. An Israeli consulting firm was entrusted with providing guidance to the Thai Department of Agricultural Extension (DOAE), with the professional backing of the Israeli Extension Service.

The extension service which has been set up under this scheme comprises an important crop protection element. In the following, the T & V approach will be characterized briefly, and the way in which crop protection advisory work has been fitted into this scheme will be appraised.

3.7.1 Essential Features of the T & V Extension Approach

The T & V system aims at the development of a national professional extension organization within the framework of a ministry or department of agriculture. It endeavours to set up a system, which through regionally and locally based extension workers reaches the majority of the nation's farmers in a regular and dependable manner.

As implied by the name, the two salient features of this approach are meticulous, continued training of extension staff at all levels, and regular visits to farmers to transmit knowhow effectively by direct contact.

Training. An essential first step has been to recruit, post and train many subject matter specialists (SMS) for general cropping advice. The SMS work out of provincial extension offices headed by an extension director. They train village-level advisers in fortnightly training sessions, with emphasis on the practices the

---

1 In cooperation with Mr. N. Minanandana and Mr. U. Dechmani, Ministry of Agriculture and Cooperation, Bangkok, Bankhen, Thailand

farmers will have to apply in major crops during the following 2-week period. The SMS frequently accompany village advisers in their field work.

In order to ensure that the SMS are in possession of the knowledge they have to transmit to village-level advisers, the SMS attend monthly workshops, usually held at experiment stations. This contributes to a welcome dialogue between extension and research. The SMS also carry out trials and demonstrations in farmers' fields.

In the past, the crop protection service had been organized within the DOAE as a separate unit, and it has remained there under the T & V system. Its specialists provide the training and support which the advisers in the field require.

Visits. Provinces and subdistricts are divided into geographically compact areas with 600–1000 farm families, and one village-level adviser is responsible for each area. The areas are further subdivided into eight groups with 75–125 farm families each; thus the village-level adviser can regularly visit each group once in 2 weeks, and he does so on a fixed day. Since not every farmer can be visited individually, nor would this be efficient, "contact farmers" are selected for intensive advisory work, and they serve as focal points for the dissemination of knowhow. About 10% of the farmers may be chosen for this purpose, and neighbouring farmers are invited to join during the adviser's visit to the contact farmer. Great care is taken in the selection of contact farmers, and various members of the farming community are usually consulted before a final choice is made. One agronomist supervises the work of eight village-level advisers.

Thailand is a large country, with 72 provinces. Thousands of village-level advisers and about one thousand supervising agronomists have been trained, together with several hundred subject matter specialists.

3.7.2 Crop Protection in Thailand

The crop protection establishment in Thailand consists of three principal groups:

a) Research is carried out by the Department of Agriculture (DOA) of the Ministry of Agriculture and Cooperatives, and by the divisions engaged in crop protection disciplines (phytopathology, entomology, etc.) at agricultural universities.
b) Extension work is done by the Department of Agricultural Extension (DOAE) at the same ministry, and specifically by the Crop Protection Service divisions.
c) The private sector comprises local manufacturers and formulators, and representatives of foreign pesticide manufacturers.

As in many developing countries, advisory units and officers fulfil many non-advisory functions, especially pest control operations, when pest or disease outbreaks affect larger areas and jeopardize yields, and when transport of infested produce from one area to another may endanger crops over larger parts of the country.

In actual practice, the time field staff actually devotes to its principal function, viz. field visits to monitor, diagnose and recommend control measures, is largely limited to the beginning of the cropping season. Later in the season, when pest populations increase and advice would be most urgently needed, the officers tend to be involved in the execution of crop protection operations, such as spraying or spreading baits to control pests affecting wider areas.

The organizational structure of the Crop Protection Division reflects the dual responsibility of the unit for both extension work and pest control. Technical policy-making is the domain of headquarters at Sankhew, where subdivisions cover the main crop protection disciplines: insect pests, plant pathology, vertebrate and weed control, with others looking after specific crops or assignments: locust control, control of sugarcane pests and diseases, beekeeping and supportive areas like biological control, plant clinics, and training centers.

There are six regions within the Department of Agricultural Extension and nine are planned in the future under the National Agricultural Extension Project; at the regional level there are Regional Plant Protection Officers and Subject Matter Specialists as well as between four and six Plant Protection Service Units per region, in charge of control operations.

The Plant Protection Service Division is relatively well endowed with pest control facilities: housing, transport, communication, storing, and pesticide application equipment. Unlike most of the extension staff, crop protection field staff reports directly to the Director of the Plant Protection Service Division at headquarters and has no formal administrative ties with regional authorities like the Regional, Provincial or District Offices of the Department of Agricultural Extension.

3.7.3 Crop Protection Advice in the Framework of T & V

The organizational set-up described above ensures coordination on the national level, but creates certain difficulties in integrating the activity of the crop protection staff with the crop husbandry extension personnel at the village level. Due to their law-enforcing duties, the crop protection officers encountered difficulties in joining the overall effort of the Department and in fully adapting to the latter's day-to-day activities, i.e. putting the T & V approach into practice.

The main bottleneck at the beginning was, however, the technical level of the crop protection advisory personnel at lower levels. To overcome this, the help of the crop protection officers, with their technical background and practical knowledge, proved invaluable. But the increasing number of village-level workers, and the regular fortnightly sessions and field visits made to train them, have made it necessary to appoint subject matter specialists in crop protection, with full-time extension duties. These specialists have joined the staff of regional offices and man the plant clinics. They thus gain first-hand experience in both field and laboratory work, and benefit from the expertise of the regional crop protection officers. The latter have also been instrumental in moulding, in a relatively short time, newly recruited university students into subject matter specialists.

Much of the training work has been done within the framework of a Thai-Israeli programme in crop protection (Ausher et al. 1982). This, in addition, has given priority to four concrete aims:

a) Cotton pest scouting.
b) Up-grading of the plant clinics.
c) Promotion of bee-keeping (which will not be dealt with here).
d) Development of new techniques in the monitoring and control of pests in fruit crops.

Cotton Pest Scouting. The area under cotton had decreased markedly in the past, owing to pest problems. To achieve better pest control at reasonable cost, a cotton pest scouting scheme was set up in various successive stages. To begin with, the problems were accurately defined at field level, and a framework for the organizational and professional aspects of the scheme was proposed. Thereafter 165 plant protection officers and village-level advisers were given practical training courses in pest identification, methods of scouting, and the practice of scouting for pests and trapping them.

This scheme has been instrumental in increasing total cotton acreage in Thailand from 24,000 ha in 1976 to almost 200,000 ha in 1981. The latter included 60,000 ha under the supervision provided by the scheme. Farmers who strictly adhered to the scheme could reduce the number of applications from 14–15 to 7–8 per season.

A joint committee for cotton pest scouting was subsequently set up, with the participation of research entomologists and extension officers, to guide future cotton scouting and suggest pest control research subjects for the following season.

Plant Clinics. The improvement of pest and disease identification has been recognized as an urgent task. A scheme has therefore been worked out, and is now in the early stages of execution, for setting up regional plant clinics, the work of which is to be professionally coordinated by a central plant clinic already established at the Department of Agricultural Extension, Bangkok.

To be integrated in the T & V and scheme, the findings of the plant clinics are brought to the attention of advisers in the bi-weekly sessions, and of SMS and research workers at the monthly workshops. Clinic staff accompanies SMS in their field visits and take part in their field trials.

Monitoring and Control of Pests in Fruit Crops. The total area of fruit amounts to about 1 million ha, but pests are controlled properly only in large commercial orchards, not in small, mixed plantations. Work at present is focussing on two major pests: (a) Fruit flies cause enormous damage to a wide range of fruits; cover spraying gives effective, but costly control. Control by the use of hydrolyzed protein baits is now being tested as a method easily employed by small farmers. (b) The mango hopper (*Idiocerus niveosparsus*) devastates mango orchards; weekly applications of spray are advocated, but rarely given by small farmers. A preliminary study has shown yellow sticky traps as a useful guide to time pesticide applications. Tests on a larger scale are to establish economic thresholds for spraying and eventually prevent excessive or wrongly timed use of pesticides.

3.7.4 Interrelations Between Extension and Research: Past, Present and Future

Two of the principal spheres, in which extension and research meet, are the formulation of pest control recommendations and the execution of a well coordinated programme for field trials.

Pest Control Recommendations. These are discussed in and issued by National Pest Control Committees, which are chaired by representatives of research institutes. These countrywide recommendations – in a country as large as Thailand – require adaptation to local conditions. This is increasingly being done in monthly workshops held in many regions all over the country, where research workers, crop protection officers and subject matter specialists working on a given crop come together at regional research stations. When research results become available, drawing prompt conclusions from them and bringing them to the farmer's attention is a major test of an efficient extension organization. The subject matter specialists consider this one of their foremost tasks.

Coordinated Field Trials. To adapt country-wide recommendations to local soil, weather and cropping conditions, field trials are indispensible. Subject matter specialists in crop protection are increasingly involved in such trials in all their stages: planning, execution, analysis, reporting, and practical application of results. Presentations of findings made in field trials are made in the monthly workshops and in the annual meetings of the recommendation committees.

3.7.5 A Final Word

The adoption of the Training and Visit extension approach by the Thai Ministry of Agriculture and Cooperatives in 1977 envisaged gradual expansion of the system to cover the whole country. In fact, by 1982, all 72 provinces of Thailand were following the T & V system, and by that time those provinces which had been first to adopt it were already reaping the fruits of several years' experience.

The emphasis initially laid on developing simple technological messages to which each farmer can relate at his own level, has proved right to get the scheme started: extension work concentrated on improvement of varieties, and basic farm practices, such as getting better crop stands and controlling weeds, all this on a number of selected crops only. Meaningful results have thus been achieved in rice production, and in the south also in rubber plantations.

Crop protection problems proved from the outset to be major topics in the biweekly training sessions. Fortunately, as mentioned earlier, a body of 25 experienced plant protection officers attended many of the training sessions, and shared their knowhow with the newly recruited extension workers. However, since these officers carry numerous responsibilities in addition to extension work, the need for training competent subject matter specialists in crop protection was accorded high priority, and this takes time.

At present, at the *provincial* level, SMS are in action only a advisers for field and fruit crops, while the service to be given by SMS in crop protection is still being provided by the plant protection officers, whose number is too limited (two or three for each province). But at the *regional* level, with each region comprising 10–15 provinces, SMS for intercrop disciplines have been appointed, especially for water management, farm management, farm mechanization and, so far in only a few regions, crop protection.

The first steps in applying the T & V system to cropping have improved the use of inputs and have lead to better farm management.

Greater intensity in farming has, as was to be expected, lead to new and more severe problems in crop protection. This has demonstrated the central role to be played by crop protection advice in the whole scheme of advisory work. It has also shown the supreme importance of having crop protection recommendations formulated on the basis of the best and most up-to-date information available and with the participation of all concerned: research staff, extension workers and farmers, who should hammer out recommendations in their workshops and periodical meetings.

References

Ausher R, Golan Y, Minanandana N, Shoham Ch, Stern Y, Dechmani U (1982) The joint Thai-Israeli programme in plant protection. Minist Agric Coop, Dep Agric Ext, Bangkok, Thailand (mimeographed)

Benor D, Baxter M (1984) Training and visit extension. World Bank, Washington DC

Benor D, Harrison JQ, Baxter M (1984) Agricultural extension. The training and visit system. World Bank, Washington DC

3.8 Advice on Pest and Disease Management: a Diversity of Objectives and Approaches

J. PALTI and R. AUSHER

The scope of advisory work in pest and disease management, its professional, legal and organizational framework, and the supporting services it relies on, clearly comprise a great diversity of objectives and approaches. The differences are most pronounced where such work in developed farm economies is compared with that in less developed ones. In considering how far the advisory systems evolved answer the needs for pest control advice, and what developments are to be desired, we shall unavoidably repeat some of the material and views presented in earlier chapters, especially in Chapter 2.6.

The main objective of crop protection in countries with advanced farm economies is the prevention of the reduction of farm income by the action of harmful biotic or abiotic agencies, so that maximum farm revenue can be achieved at reasonable risk. In developing farm economies, crop protection objectives are different: the element of risk looms large in countries in which most crops are grown for subsistence, epidemics imply hunger, and no reserves for low-yield years are available. Risk considerations are slightly less pronounced where a smaller or larger part of the crops grown, e.g. cotton, coffee etc., is for sale, and crop failure does not mean immediate shortage of food.

The objectives of crop protection everywhere depend principally on the interaction of the following factors:

a) The farmer's personality, which determines his attitude to risk, and his willingness to cooperate and to follow advice.
b) The farmer's educational level, his know-how, and the availability of bodies that generate and disseminate know-how.
c) The farmer's resources and his readiness to commit them to the protection of his crops.

In the following we shall distinguish between the desiderata of advice for crop pest and disease management in countries at various stages of development, and in conclusion consider some aspects relevant to all such stages.

3.8.1 Advanced Farm Economies

The methods adopted to dispense crop protection advice in countries with farm economies at comparably high levels vary greatly. A partial analysis of the reasons underlying this variety yields the following points.

Intensity of Advice as Related to Number and Specialization of Growers. There are extreme differences in the number of farm units and the average size of farms served by advisers in various countries. Thus, in the Federal Republic of Germany (FRG), there are several hundreds of thousands of farmsteads on about 12 million ha (average farm size less than 25 ha), whereas in California 82,000 owners and operators farm 4 million irrigated and 11.8 million non-irrigated land, and the (somewhat theoretical) farm size is 10–20 times as large as in the FRG. This clearly affects the intensity of extension work, as expressed by the number of farm families served by one adviser. This aspect may be somewhat less important in crop protection than in general cropping advice, because crop protection advice is often given indirectly, i.e. to the crop specialists who then transmit it to the farmer, less frequently straight to the farmer. Intensity of advice also relates to the manner in which the farmer is contacted: the greater the number of farmers to be served, the more important the role which mass media tend to assume in disseminating advice. The level of growers' specialisation is also an important aspect: specialised growers are mostly familiar with routine crop protection matters, and need advice on more complex problems. Since the adviser needs time to study such problems, the number of specialised growers he can directly advise is necessarily limited.

Growers' Participation and the Target Groups of Advice. An important difference between the services given in various countries lies in the degree to which growers participate in running, and even financing, them. Texas seems to stand out in the involvement of growers and their associations in the active operation of integrated pest management (see Chap. 3.6.3). In England, the Federal Republic of Germany and Israel, growers influence the policy of crop protection advisory work, but play a minor role in its day-to-day operation.

The target groups of crop protection advice in many states of the U.S.A. include a large and ever-growing number of private consultants. In California alone, there are at present 5800 licensed, privately employed pest control advisers, as well as about 10,000 applicators. The advanced educational level of the consultants may be expected to have two main consequences: (1) The public services will be required to make available a current supply of data (weather, regional pest and disease distribution, prices etc.) which serve the consultant to make his own decisions. (2) The public adviser, if he wants to be more than a supplier of data, will have to advance his own professional level, and be increasingly involved in local experimentation, to furnish specific, up-to-date advice. The role played by private consultants in European countries is as yet limited, but may also be expected to gain in importance in the near future. The size of the advisory staff of agricultural suppliers, especially pesticide manufacturers, in many countries far exceeds that of public agencies, as has been discussed in Chapter 2.6.2. Like private consultants, they represent an important target group for advice, and they also generate essential information on control by pesticides.

University-Linked Compared with State-Run Services: Generation and Validation of Know-How. In many states of the U.S., extension specialists are located in universities, and thus have a direct connection with research facilities. In most other countries, including Western Europe and Israel, advisory services are run by the state (see also Sect. 2.6.1.6). The extent to which these services are involved in research differs widely. In England and Wales, ADAS has its own laboratory facilities not only in the six regional centres of advisory work, but also in four central laboratories, each of which deals with another aspect of crop protection (see Table 1 in Chap. 3.1.2). In small countries, such as Israel, separate advisory laboratories are not required, since most advisers have ready access to research stations.

The generation of know-how and technology, which is so essential for extension services, involves several steps:

a) Identification of production constraints, with subsequent determination of priorities in Research and Development (R & D) and of the intensity of the work required to solve problems.
b) Definition of the most appropriate channels for R & D work, which generally comprises (1) field trials by advisory personnel, often in farmers' fields or orchards, and (2) selection of the research facility best suited to carry out the research required, and provision of resources adequate for obtaining results with reasonable speed.
c) Adaptation of research results to a wide spectrum of field conditions and integration of these results in the production system.

In crop protection, the identification of constraints must be based on a comprehensive appreciation of the crop in health and ill-health, and on up-to-date data on the economics of crops, of pest damage, and of pest management expenses. This overall appraisal can in many cases better be made by the crop-oriented staff of state-run advisory services than by research staff specialised by crop protection disciplines. The participation of growers in the appraisal process is always desirable. However, advance in agricultural research and technology greatly depends on awareness and availability of developments in other fields, and here the university-linked system may have an advantage.

In the performance of problem-oriented R & D work, the first step is often the execution of on-farm field trials, which may either solve the problem or crystallize specific need for further research. Each adviser should have an adequate amount of his time allotted to such trials, and this is to be considered an important component of his professional status. On-farm trials cost very little, and have the added advantage of getting farmers involved in the development of new technologies. In practice, the phase of field trials and research often overlap, which can be of mutual benefit.

In state-run services, the allocation of funds for research depends primarily on government budgets and on growers' pressure, and the research institution that seems most suitable can be chosen to perform the research. In university-linked services, research is usually performed at the university concerned. Here additional academic considerations also play a part in fixing priorities and funding research, e.g. the frequently encountered inclination for single-discipline rather than multidisciplinary research, and the relevance of the research proposed to the academic interests of the staff.

The third step, adaptation of experimental findings to diverse conditions and their translation into practice, requires a competent staff able and willing to spend most of its time in the field. This involves repetitive tests under a variety of soil, weather and cultural conditions. The various levels of advisory personnel in state-run services are best suited to this task. Moreover, where practical application of research findings is of immediate urgency, as is often the case in crop protection, state-run services can draw on a sizeable pool of professional personnel.

The integration of findings in any sphere of agricultural research into the whole production system is often the final and decisive step in the usage of research results. In many cases, crop protection must adapt itself to the adoption of practices developed to increase farm income, but with incidental effects on certain pests and diseases. A case in point is the introduction of trickle irrigation for industrial tomatoes in Israel: compared to overhead irrigated fields, those under trickle irrigation proved markedly more susceptible to the attack of red spider mites and of powdery mildew (*Leveillula taurica*), but less liable to attack by late blight (*Phytophthora infestans*). This caused a revolution in crop protection measures applied to such tomato crops. Conversely, introduction of disease-resistant varieties in many crops has to be followed by adaptation of sowing, fertilizing and irrigation practices to the requirements of these varieties. Advancement of crop production as a whole by integration of such specific research findings, would appear to be easier in state-run advisory services.

In sum, the important point appears to be that the crop protection advisory service should be crop- and grower-oriented, and not specifically pest- or disease-oriented. This facilitates contact with farmers and crop specialists, and affords the adviser a comprehensive and integrated approach to his work.

3.8.2 Farm Economies in Developing Countries

Farming communities in developing countries may be divided, for the purpose of crop protection advice, into several categories:

a) Subsistence farmers, largely illiterate or semi-literate.
b) Farmers growing crops for sale through individual marketing, most farmers semi-literate or literate.
c) Farmers marketing crops through cooperatives or government agencies; most farmers semi-literate or literate.

This classification, though arbitrary, may serve to clarify several points.

Matteson et al. (1984) have rightly pointed out that understanding traditional agricultural systems is the starting point for farming development, and this decidedly includes the study of the use of traditional pest management systems. Getting farmers acquainted with changes is best effected by visits of advisers to their fields: however, this work has become easier where even illiterate farmers have access to mass media, mainly radio; it appears that, once you get to them, many traditional farmers are much less conservative than has generally been assumed.

In crop protection work in general, and most particularly in developing farm economies, the farmer's trust in the advice he is given is of the utmost importance. It often is a hard struggle to obtain his trust and, once obtained, it must not be endangered. Therefore, the *reliability* of the advice given may have to be the central consideration and failures cannot be risked. This sometimes entails advice for chemical pest and disease control, as a measure which is safe, although it may not be absolutely necessary.

In all the above three categories of agricultural communities, possibilities for cooperation exist, but they are much better when backed by structured cooperatives, e.g. for water distribution, or by marketing organizations which have the resources to support communal activities.

Some of the tasks facing the crop protection adviser in communities within the three categories are discussed in the following sections.

Subsistence Farmers. Subsistence farmers practice polyculture or shifting cultivation in small plots. As stated elsewhere, their main concern is to feed their families until the next crop ripens: they do not aim at maximal yields of any crop, but at a steady and adequate supply of food throughout all seasons. Since crops are not sold, no cash is available for buying pesticides or equipment for their application.

Pest and disease populations in such crops are largely endemic (Putter 1980), and their development is limited by natural enemies of the pests, and by resistance to parasites in hosts grown traditionally over many generations. Epidemics break out occasionally, mostly triggered by unusual weather or due to sudden influx of swarms of pests.

Under these conditions, the function of the crop protection adviser is chiefly to look out for anything that could upset the equilibrium between endemic parasites and their hosts. A primary task is to encourage sanitation, e.g. removal of leaves or fruits affected at an early stage of crop growth, or collection of harmful insects by hand, and also destruction of crop residues soon after the harvest. Yields can be improved by gradual introduction of more resistant varieties and of better-quality, disease-free and properly disinfected propagation material of traditional crops.

Individual Farmers Growing and Marketing Their Crops. Where crops are sold on the market, some money is likely to be available to the grower, if he so desires, to buy equipment and materials for crop protection on a small scale. Availability of the right pesticide at the right time will often be a problem (Chap. 3.6.2). Equipment may be given to the farmer by the government either without charge or on hire.

But even if pesticides and equipment are available, the approach of the farmer to their use may be dictated primarily by consideration of the labour involved in application and of the alternative use of his limited resources (see Chap. 2.3.8) or by sheer ignorance of potential benefits. Here the task of the adviser may be to help the farmers in weighing the practical pros and cons of pesticide use, introducing considerations of the price of produce and pesticide, or of the potential danger to crops, with which the farmer is not fully familiar.

Here again, importance of sanitation must be stressed. Where the sizes of the plots of crops to be marketed tend to become larger, the advantage of intercropping or polyculture in restricting pest development must not be lost sight of.

Farmers Growing Crops for Cooperative or Government Marketing. The chief difference between farmers of this group and those mentioned above is, firstly, that the price paid for the produce may be fixed, especially for export crops like cotton, groundnuts or coffee, and the grower can thus estimate how much he can afford to invest in crop protection; and secondly, that pesticides and application equipment are often made available by the marketing agency, or the cooperative. These bodies could also provide crop protection advice, by means of an instructor who may or may not be a crop specialist.

Crops grown under the aegis of governmental or other public agencies often tend to be larger in extent than plots grown by farmers mentioned in the first two categories. They may lose some of the protection from pests that beneficial insects afford to mixed crops. Pest control by pesticides then becomes increasingly important not only in individual plots but also in those in the immediate vicinity. The task of the crop protection adviser then includes demonstration of the advantages of simultaneous pest control in groups of plots or in all the plots of the endangered crop.

Scope and Structure of Crop Protection Advice in Developing Farm Economies. As mentioned in Section 2.6.1.2, crop protection advice in many developing countries is being given within the framework of government agencies which also fulfil regulatory and law-enforcing functions. This neither allows the adviser sufficient time for regular field visits nor for keeping abreast of current technical development and, perhaps worst of all, makes it hard for him to gain the farmer's trust. Greatly preferable is a system which emphasizes the professional component in the adviser's work; its aim is to make him into a crop protection specialist and involve him in the process of generating technology, with a well-planned working relationship with local research. He then plays an important part in pin-pointing the farmer's needs and in formulating relevant technical messages. Freed from regulatory duties, the adviser then enjoys regular training and up-dating of information, and contributes his information based on systematic field visiting schedules. This approach to the development of higher professional levels of advisers, the Training and Visit system has been described in Chapter 3.7. This system is now being applied in about 40 countries, mainly in Asia and Latin America, and lately also in Africa. Results obtained after a few years show that the proficiency of crop protection advisers, from the village level upwards, can be effectively improved by this system of intensive staff training and frequent well-planned visiting schedules, covering all of the farming communities in the region.

3.8.3 Farm Economies at All Levels of Development: Subjects That Require Special Attention

A number of subjects that require special attention, at all levels of farming, and that have not so far been given proper emphasis, will here be mentioned in conclusion.

Determination of Treatment Thresholds. This is an essential tool of Integrated Pest Management and has been discussed in Chapter 2.3. Since thresholds depend on the integration of so great a variety of weather, crop, pest and pathogen, and market factors, their determination has become much enhanced with the development of computer models. But these models are only as good as the data on which they are based; and lack of reliable data, more than anything else, delays development, validation and application of many more models. Here the crop protection adviser can be of the greatest help: in many cases, nobody else is so well-placed to supply large bodies of field data on crop growth and cultural practices, and on pest characteristics, such as food preference, destructiveness at each developmental state, life expectancy, generation cycle etc. Wherever possible, part of the crop protection adviser's time should deliberately be reserved for the collection of such data, and for participation in research for the determination of treatment thresholds.

Pesticide Application Techniques. Pesticide application, a subject treated in some detail in Chapter 2.5, has received only cursory attention in the desriptions of advisory schemes for crop protection which have been presented in Chapters 3.1 to 3.7. This is hardly surprising: the volume of research into application techniques is so small – and its size is so disproportionate compared to that of pesticide research – that there simply is not enough know-how reliable enough to give farmers responsible advice. A few of the larger manufacturers of pesticides have recently begun to make some investigations, especially in relation to low volume application techniques. It is greatly to be hoped that more commercial undertakings will contribute to this neglected aspect of crop protection as they realize that the economic use of pesticides and avoidance of ecological hazards depends so much on the ways in which pesticides are used.

Biological Control of Pests and Pathogens. Great advances have been made in various high-level farm economies in studying the fauna of predators and hyperparasites of economic pests. The result has been a mixture of impressive successes and fairly frequent failures. At times, initial successes have turned into failures because the advent of new pests has changed the picture. A case in point is the biological control of citrus scale insects in Israel: The introduction of hyperparasites from Asia was extremely successful for a considerable number of years, but has lost some of its reliability with the spread of an additional scale not controlled by these hyperparasites. The lesson to be learned from this and similar occurrences elsewhere is that biological control of pests, which is so important a component in Integrated Pest Management, requires constant close supervision to ensure early detection of any interference in the desired pest control. This frequently is not a task the crop protection adviser can handle himself, but he can make fundamental contributions to the training of supervisors employed for this purpose.

The study of biological pest control agents in developing farm economies is patchy and, with rare exceptions, has not yet been of practical significance. It needs a great deal of research in the performance of which the adviser should play an important role.

The biological control of soil-borne diseases is receiving much attention in advanced countries. Research on this subject seems promising, but practical success is still elusive.

Cultural Practices. The effects of cultural practices on the prevention and control of pests and diseases differ greatly from region to region, and may even be field-specific. Variations in sowing dates are important measures, especially where seasons widely differing in rain and temperature succeed each other. Irrigation and water management, the cultural practices with the most far-reaching effect on diseases, can be applied in many modes and at greatly differing rates and frequencies. This is of special significance for the developing countries in which most of the lands needing irrigation are to be found. These are only a few examples, but they point to the fact that restriction of pest and disease development by cultural practices must continually be demonstrated and brought up-to-date in all advisory work on crop protection. The opportunities for doing so are numerous (Palti 1981), and many of the measures to be recommended are effective only when applied on a regional scale, with the cooperation of the farming community as a whole. The example of the rice pest *Niliparvata* may be mentioned here (Table 1), because it shows the great variety of cultural practices that may contribute to the control of this destructive pest of rice in South and South East Asia (Anonymous 1982).

Full use of all the opportunities offered by cultural measures constitutes a most important part of all efforts at Integrated Pest Management. As mentioned above, effects of pesticides are often more spectacular and predictable and may for this reason be preferred where risks of pests and diseases are to be avoided

Table 1. Cultural practices affecting the control of *Niliparvata* sp. in rice in South and South East Asia. (Anonymous 1982)

A. Practices that have a *positive* effect on control:
1. Synchronous planting
2. Synchronous harvesting
3. Drainage
4. Flooding
5. Planting mixed varieties or lines
6. Short-duration varieties

B. Practices that have a *negative* effect on control:
1. Delay in the time of transplanting
2. Direct seeding
3. High nitrogen inputs

at all cost. But the crop protection adviser must be considered remiss, if he does not persevere in his efforts to develop other practices contributing to the protection of crops.

3.8.4 The Crop Protection Adviser of the Future and His Aspirations

As in all professions, training, motivation, opportunity and the status attainable by dedicated work are the factors that make it possible to attract competent personnel to all levels of crop protection advisory work.

It appears to us, that in several ways crop protection advisers in many countries do not enjoy as much professional recognition as comparable staff in other fields of endeavour, and their status is somewhat ill-defined. We offer here a number of considerations how this status might be improved to motivate high-class personnel to take up crop protection advice, much of it at the level of field and farmer, as a rewarding profession.

Training. The comprehensive knowledge and flexibility demanded from the adviser have variously been stressed in this book. To achieve both, he should be given the chance to widen and deepen his expertise beyond what is now offered in most universities, and acquire a well-rounded view of the entire field of crop protection, and its integration with crop production systems. This involves constant refreshing and up-dating of knowledge as an integral part of his work in the advisory system.

Opportunity and Motivation. The opportunity for material advance is somewhat brighter for the crop protection adviser than for his colleagues, the crop advisers, because manufacturers of pesticides and of application equipment are inclined to employ personnel with crop protection experience gained in advisory work. But apart from entry into the commercial field, chances for formal advancement in crop protection advisory services are usually quite limited, except perhaps in the university-linked extension systems in the U.S.A. This limitation invariably affects motivation, unless the adviser is offered wide opportunities for satisfactory professional progress; if the latter presents a real challenge, it often provides the major motivation for work in the service.

References

Anonymous (1982) 11th FAO UNEP Panel on Integrated Pest Control, Koala Lumpur, Malaysia, March 1982

Matteson PC, Altieri MA, Gagna WC (1984) Modification of small farmer practices for better pest management. Annu Rev Entomol 29:383–402

Palti J (1981) Cultural practices and infectious crop diseases. Springer, Berlin Heidelberg New York, 243 pp

Putter CAJ (1980) The management of epidemic levels of endemic diseases under tropical subsistence farming conditions. In: Palti J, Kranz J (eds) Comparative epidemiology. PUDOC, Wageningen, pp 93–103

Index of Pests, Pathogens and Beneficial Organisms

Subject Index